Rudie H Kuiter: Basslets • Hamlets, and their relatives. A Comprehensive Guide to Selected Serranidae and Plesiopidae

First English language edition 2004

Published by TMC Publishing, Chorleywood, UK

Design, editing and taxonomic decisions: Rudie H. Kuiter.
Print and production: Grupo M & G Difusión S.L

TMC Publishing
Solesbridge Lane,
Chorleywood, Herts WD3 5SX
United Kingdom

Tel: +44 (0) 1923 284151 Fax: +44 (0) 1923 285840
Email: info@tmc-publishing.com
Website: www.tmc-publishing.com

ISBN 0-9539097-6-X

The Marine Fish Families Series

Basslets • Hamlets

and their relatives

A Comprehensive Guide to Selected Serranidae and Plesiopidae

Rudie H Kuiter

TMC
publishing

TMC Publishing, Chorleywood, UK

Acknowledgements. Due to the large number of genera and species, many of which are rare or rarely noticed by divers, other than those that take a personal interest in particular groups, this volume could not have been completed to the standard required for the series without the contributions of many photographers. I would like to thank the regular contributors who generously provided photographs and advice: Jerry Allen, Helmut Debelius, Ken Graham, Roger Steene, Phil Heemstra, Rick Winterbottom, Scott Michael, Paul Humann, Ross Robertson, Gary Bell, Malcolm Francis, Thomas Paulus, Phil Woodhead, Tomonori Hirata, Akira Ogawa, Hiroyuki Tanaka, Osamu Hoshino, Yasuaki Miyamoto, Satoshi Yamamoto, Shoichi Kato, Masatoshi Kubo, Robert Myers, Patrick Louisy, Neville Coleman, Frank Schneidewind, Peter Wirtz, and many others. All are credited in the caption with their photographs.

TABLE OF CONTENTS

INTRODUCTION

There are many small, tropical, and often colourful percoid fishes, commonly referred to as 'basslets' found on reefs that belong to a number of families. 'Bass' are the larger percoid species, including the groupers, and the more edible fish, and the names are used for species belonging to a number of different families. The majority belongs to the large family Serranidae and this comprises several subfamilies, and in general it is the subfamilies with the smaller species that are comprised of the basslets or dwarf seabasses. Some, going under similar common names, are placed in separate, but closely related families. In all, this book deals with the majority of such species and included are the anthiids, generally known as fairy-basslets or basslets, sea-perches or perches; the liopropomids, reef-basslets or bass; grammistiids, the soapfishes; the subfamily Serraninae with its dwarf seabasses and hamlets; Grammatidae, also known as basslets; and the plesiopids that are comprised of several distinct groups such as longfins, roundheads or devil- and hulafishes. Most of these fishes feed on small swimming or crawling prey, usually remain small when fully grown, and often are exceptionally colourful. However, a few grow to medium size and some related species living in cool waters can reach about 50 cm in length. The smaller, and usually the more colourful, species are ideal candidates for the reef-aquarium that supports living corals or other various sessile invertebrates, systems that have recently become very popular. The larger species are best kept in public aquariums where space is not a problem.

Pseudochromids or dottybacks, also ideally suited for the reef-aquarium, could be considered for this volume. They are basslet-like, but, as there are over 200 species, these will be treated in a separate volume.

Amongst the various groups, there are a number of species that are known only from very deep water, too deep to dive with conventional scuba-diving equipment. Sometimes such fishes are trawled and successfully placed in an aquarium, but in most cases are rarely seen alive. I have included many deep water species, especially when other members of the same species occur at shallower depths and complete the genus, but also those often photographed during scientific expeditions. Some of the rarely seen cryptic species, commonly collected during scientific work, are also included with specimen photographs, but when ever possible, underwater pictures taken in the wild are used in preference over aquarium or specimen shots, or in conjunction with these to show certain details.

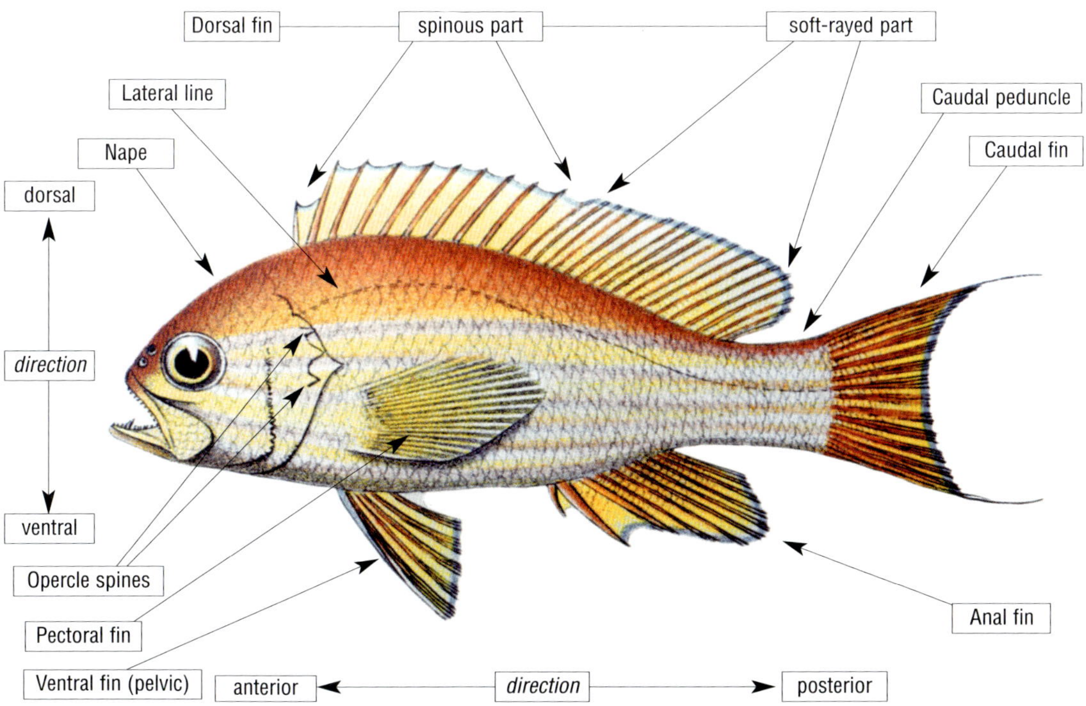

Fig. 1. Selected features of seaperches.

ABOUT THIS BOOK

This book is the 6th volume of *The Marine Fish Families Series*, dealing with species on a global scale. The aim is to create a series of books that comprehensively covers closely related or certain groups of fish with photographs, as a pictorial guide, primarily for the purpose of identifying species and to find out what is known about them. A volume may comprise a group, or related groups of fishes that belong in a single family, several families, or just part of a family, depending on how small or large such a group is. The first five volumes dealt with the orders of Syngnathiformes, Acanthuroidei, and the families Labridae (in part, whilst the other part of Labridae is planned as a separate volume), Chaetodontidae and Pomacanthidae, numbers 1–5 respectively.

Due to the taxonomic make-up of small basslet-like fishes and their relatives, this volume has a mix of groups that belong to a number of closely related families, whilst the largest family Serranidae is represented in part by the appropriate subfamilies. The main groups featured are the Anthiinae or fairy basslets and relatives, and the various similar subfamilies within the Serranidae that are comprised of mainly small fishes and are commonly referred to as 'basslets' and 'hamlets'. The majority of the larger species within the Serranidae belong to the subfamily Epinephelinae, which includes the giant groupers and coral-trout or coral-cods, comprising many of the species that are primarily of interest in the fishing industry and could fill a volume in this series in its own right. The genera that are included in this volume are treated as comprehensively as possible, with photographs of different stages, variations and representations from different geographical zones.

NAMES

Each species is presented with its current common and scientific name, followed by its original scientific name with the author and date of its description. The scientific name of a species is binominal. The first part is the genus, which be shared by closely related species, and the second part is the species name, which is unique. The common and scientific names are presented in a coloured box, separate above the main species text, and is immediately recognisable on the page. The text for each species begins with the original name (often species were first described in a different genus), its author and date when it was first described. A species may have been described and placed in the genus that was thought to be correct at the time, but other species that belong in the same genus may have been placed in a different genus. When they belong to the same genus, the oldest name is valid only and the later ones used become junior synonyms. If the same name is already in use for other taxa in the animal world, at the same level, such as a genus, family, etc, the next available name by date is used. Some of the included species may have been discovered only recently and are not named scientifically. These are placed in the appropriate genus and presented here as 'sp', short for a species (spp is plural). When the same species was described by different authors or the same author described different stages of the same species, resulting in multiple names for the same species, again the first available name applies.

Some species may be known under several common names which may reflect different forms, such as during growth or between the sexes, their use in different countries or in the same country in different interest groups. Divers may call a species one name and aquarists may use another for the same species, whilst fish-exporters often come up with a fancy name that may attract more buyers. Commercial fisherman may use yet another name on fish markets to make them more appealing to a different audience, whilst many closely related species are traded under a single name. Scientists often translate the meaning of their scientific name, which often makes no sense as this usually refers to a diagnostic character and can be very confusing for the layman. Using scientific names as actual common names is even more confusing and is not encouraged in this series of books. For example, *Chromis bicolor* becomes 'Bicolor Chromis' so why bother with a common name at all! In this book, usually a single name is given for a species. If there is a second, often used and well-established name, it may be included in the text under *remarks*. Scientific names help to make sure that we are talking internationally about the same taxon, but in some groups the taxonomy is badly in need of work and names are changing regularly on different taxon levels. In some cases a common name carries more weight than its scientific counterpart.

Compared to many other fishes, the taxonomy of the family Serranidae is reasonably well established, but primarily on a species level. Some genera are in need of revision, and in some cases there are some problems with the assignment of certain species to a genus. Amongst ichthyologists, the principal disagreements are usually with regards to applying a subspecies or subgenus level, 'where to draw the line' so to speak. This has led to much inconsistency between workers, as those most familiar with 'its' group recognises species correctly whilst those less familiar with the same group are often sceptical with regards to siblings. Often species are lumped together because less informed scientists simply can not tell them apart, but divers and aquarists often find that behaviour plays an additional role in recognising different species.

Family Contents

FAMILY SERRANIDAE - Subfamily Contents

FAMILY SYMPHYSANODONTIDAE

Comprises a single genus with 6 species. Very similar to anthiids, but some skeleton features resemble Lutjanidae.

GENUS *Symphysanodon* Bleeker, 1878

Masculine. Type species: *Symphysanodus typus* Bleeker, 1878. Comprises 6 small species, 2 of which are in the Atlantic and 4 in the Indo-Pacific. Mostly deep water species, usually living at depths of several hundred metres.

1 *Symphysanodon andersoni* Kotthaus, 1974	Arabian Sea
2 *Symphysanodon berryi* Anderson, 1970	Bahamas
3 *Symphysanodon katayamai* Anderson, 1970	Philippines
4 *Symphysanodon maunaloae* Anderson, 1970	Hawaiian Islands
5 *Symphysanodon octoactinus* Anderson, 1970	West Indies
6 *Symphysanodon typus* Bleeker, 1878	New Guinea

S. katayamai. Izu, Japan. Depth 45 m. Length 18 cm. Shouichi Kato.

Fork-tailed Seaperch
Symphysanodon katayamai

Symphysanodon katayamai Anderson, 1970. Philippines.

Southern Japanese waters and Taiwan to Philippines. Deep coastal rocky reefs, sometimes seen in relatively shallow depths in pursuit of zooplankton. Identified by the long forked caudal fin and broad yellow to orange mid-lateral stripe. Length to 24 cm including caudal fin to tips.

FAMILY CALLANTHIIDAE

A small family with 2 genera and 11 species. Closely related to Serranidae and superficially very similar to anthiids. They feature a lateral line that runs very high over the sides, just below the dorsal-fin base, have uniquely shaped nasal organs, and rows of scales mid-laterally bear a series of pits or grooves. They are generally deep-water species and only few species are known from depths shallower than 100 m, and even those are only observed in the shallower depths at certain localities. Like many deep-water species, their colouration is predominantly red with yellow.

The species which have been observed occur in schools and feed in open water on zooplankton. They are an attractive group to keep in captivity, but require regular feeding. It is essential to have live food available all day, such as mysids, or they will lose condition quickly. Most species attain 30 cm or more in length and in addition, require relatively large open spaces that usually only public aquariums can offer.

GENUS *Grammatonotus* Gilbert, 1905

Masculine. Type species: *Grammatonotus laysanus* Gilbert, 1905. Comprises 4 small species, usually living in a depth range from 100–300 m. Features are moderately large scales; short lateral line, ending below soft dorsal; caudal fin truncate or slightly concave with many rays ending with short filaments; and ventral fin positioned anterior to pectoral fin. Only *G. suragaensis* occurs at shallower depths and has been photographed as shallow as 60 m.

1 *Grammatonotus crosnieri* (Fourmanoir, 1981)	Philippines
2 *Grammatonotus laysanus* Gilbert, 1905	Hawaiian Islands
3 *Grammatonotus macrophthalmus* Katayama *et al*, 1980	Japan
4 *Grammatonotus surugaensis* Katayama *et al*, 1980	Japan

Pygmy Splendid Perch
Grammatonotus surugaensis

Grammatonotus surugaensis Katayama, Yamakawa & Suzuki, 1980. Off Numazu, Suruga Bay, Japan.

Southern Japanese waters, known from Suruga and Sagami Bay, the Izu and Boso Peninsula region, and Osumi Strait. Deep rocky reef to at least 120 m. Identified by its unusually-shaped caudal fin with thread-like filaments in posterior border. A small species, reaching about 11 cm.

G. surugaensis. Oshima, Japan. Depth 75 m. Junichi Todate.

Genus *Callanthias* Lowe, 1839

Masculine. Type species: *Callanthias paradiceus* Lowe, 1839 (=*Callanthias ruber* Rafinesque, 1810). Comprises at least 7 species, but most that are presently included in the genus are quite different in morphology from the type-species and need further investigation. The type from Europe has long pointed median fins with filamentous tips. Many other species have rounded dorsal and anal fins. All species live at moderate depths.

1 *Callanthias allporti* Günther, 1876	Tasmania	p11	
2 *Callanthias australis* Ogilby, 1899	Sydney	8	
3 *Callanthias japonicus* Franz, 1910	Japan	10	
4 *Callanthias legras* Smith, 1948	South Africa	7	
5 *Callanthias parini* Anderson & Johnson, 1984	Se. Pacific	n.i.	
6 *Callanthias platei* Steindachner, 1898	Juan Fernandez Is	10	
7 *Callanthias ruber* (Rafinesque, 1810)	Sicily	7	

n.i. = not included

Parrot Seaperch *Callanthias ruber*

Lepimphis ruber Rafinesque, 1810. Palermo, Sicily.

Widespread in the Mediterranean Sea and the east Atlantic from just south of the English Channel to Mauritania and probably most of the offshore islands. A rarely seen species that occurs in deep water near reefs, usually in depths well over 60 m, and ranging to at least 300 m. Has greatly extended caudal-fin tips and is unusual in having the longest dorsal-fin rays in the centre of the soft part, forming an obvious tip. Length to about 30 cm, excluding filamentous tips.

C. ruber. Type-species. From off Italian coast. Francesco Costa.

African Splendid Perch *Callanthias legras*

Callanthias legras Smith, 1948.
Off Algoa Bay, South Africa.

Western Cape Province to Natal, South Africa. Rarely collected species known only from deep water offshore. Occurs over rocky bottoms to a depth of about 400 m. Like its congeners, colours are a mix of orange and yellow, but the ventral fins are conspicuously white. Median fins with filamentous corners posteriorly, probably longest in males. Length to about 25 cm.

C. legras. South Africa. Phil Heemstra.

A

C. australis. Bicheno, Tasmania. Depth 35 m. Length 30 cm.

Australian Splendid Perch *Callanthias australis*

Callanthias australis Ogilby, 1899. Sydney, Australia.

Eastern Australia from central NSW to Tasmania and south coast to Western Australia, northern New Zealand and Kermadec Islands. Offshore reefs and islands, usually at depths over 25 m and often abundant on reefs that level out at 50 m. Reported to a depth of 200 m. Juveniles pink with subtle mauve markings. Females yellow to orange or pink. Males variable with mood, sometimes showing almost white fins. Length to 30 cm.

Remarks: juveniles were kept with little success, despite having live mysids available on a regular basis.

B

C

C. australis. NSW, Australia. Depth 25 m. **B** Montague I. Female. Length ~25 cm. **C** juvenile. Seal Rocks. Depth 10 m. L. 35 mm.

D

C. australis. Male in cave. Montague I., NSW, Australia. Depth 25 m.

E

C. australis. Male, unusual deep water colouration. Off Bate Bay, NSW, Australia. Depth ~120 m. Ken Graham.

F

G

H

C. australis. Montague I., NSW, Australia. Depth 20–25 m. **F** & **G** juveniles, lengths ~9–12 cm. **H** males in large cave.

Japanese Splendid Perch
Callanthias japonicus

Callanthias japonicus Franz, 1910. Japan.

Subtropical Japanese waters and China Sea. A separate population off north-western Australia appears to be this species. Deep rocky reefs, usually in excess of 60 m. Forms small aggregations when feeding in open water on zooplankton. Often mixes with other feeding planktivores, especially anthiid species such as *Sacura*. Pink to orange yellow with yellow to orange fins. Length 25 cm, excluding fin-filaments.

C. japonicus. Tokai Aquarium, Japan. **A** & **B** male. **C** female. Lengths about 22–25 cm.

Chilean Splendid Perch
Callanthias platei

Callanthias platei Steindachner, 1898. Juan Fernandez Islands, Chile.

Known from the Juan Fernandez Archipelago where common around the islands. Usually seen at depths over 25 m, near large caves and overhangs. Swims in open water when feeding on zooplankton, often in schools mixed with an apparently undescribed *Caprodon* species. Females mainly orange and males mauve. Develops long filaments on caudal fin tips. Length to 30 cm.

C. platei. Robinson Crusoe I., Juan Fernandez Is., Chile. Depth ~25 m. Lengths about 20–28 cm. Pedro Niada.

Southern Splendid Perch *Callanthias allporti*

Callanthias allporti Ogilby, 1899. Tasmania.

A temperate species known from southern New Zealand and Tasmania, Australia. Known only from trawled specimens in deep water in Australia, but can be found as shallow as 25 m or less in Fiordland, southern New Zealand. Rises above the bottom to feed on zooplankton, usually forming small aggregations, or mixes with the more common butterflyfish *Caesioperca lepidoptera* when at shallower depths. Plain pink with a yellowish head and ventral fins. Adults often develop long filaments on the caudal fin tips. Length to 26 cm, excluding filamentous fin tips.

A

C

B

C. allporti. Fiordland, New Zealand. **A** & **B** Malcolm Francis.

D

E

C. allporti. Fiordland, New Zealand. **C–E** Quentin Bennett.

Anthias **14**

Holanthias **17**

Hemanthias **18**

Hemanthias **18**

Pronotogrammus **19**

Pseudanthias **20**

Nemanthias **82**

Rabaulichthys **84**

Tosana **85**

Luzonichthys **86**

Tosanoides **90**

Odontanthias **92**

Sacura **96**

Serranocirrhites **98**

Caprodon **99**

Lepidoperca **102**

Caesioperca **104**

Plectranthias **108**

Othos **114**

Hypoplectrodes **115**

Acanthistius **120**

Epinephelides **124**

Trachypoma **125**

SUBFAMILY ANTHIINAE, GENUS *Anthias* Bloch, 1792

Masculine. Type species: *Labrus anthias* Linnaeus 1758. Primarily an Atlantic genus with at least 7 species, plus 1 in the eastern Pacific. All live in deep water over rocky substrates, mostly around sea-mounts or rock ridges that are swept by strong currents, where they form schools and feed on zooplankton. Most live in depths over 100 m and only one species, commonly seen by scuba divers in the Mediterranean and eastern Atlantic. This species is rarely kept in the home aquarium as they require cool conditions, lots of space, and regular feeding. Due to these requirements, they are best kept in public aquariums.

1	*Anthias anthias* (Linnaeus, 1758)	Mediterranean Sea and eastern Atlantic	16
2	*Anthias* cf *anthias*	Cape Verde, eastern Atlantic	16
3	*Anthias asperilinguis* Günther, 1859	South America, south-western Atlantic	n.i.
4	*Anthias helenensis* Katayama & Amaoka, 1986	North of St. Helena I., south Atlantic	n.i.
5	*Anthias menezesi* Anderson & Heemstra, 1980	South America, south-western Atlantic	n.i.
6	*Anthias nicholsi* Firth, 1933	North-western Atlantic	17
7	*Anthias noeli* Anderson & Baldwin, 2000	Galapagos I., eastern Pacific	n.i.
8	*Anthias salmopunctatus* Lubbock & Edwards, 1981	St. Paul's Rocks, mid-Atlantic	n.i.
9	*Anthias tenuis* Nichols, 1920	Bermuda, Atlantic	17
10	*Anthias woodsi* Anderson & Heemstra, 1980	North-western Atlantic	n.i.

n.i. = not included

A. anthias. Male? Variation with pointed ventrals and distinctive colouration. Azores. Depth 55 m. Peter Wirtz.

Anthias anthias. Above: schooling and feeding. A hogfish *Bodianus scrofa* near the bottom. Azores. Peter Wirtz.
Below: males. Aquarium, Mediterranean L. 17–22 cm. Patrick Louisy.

Mediterranean Basslet
Anthias anthias

Labrus anthias Linnaeus, 1758.
Southern Europe.

Widespread Mediterranean and Atlantic from Portugal to the Azores and Canary Islands, and Angola (if one species). Occurs mainly on deep reefs with caves, large boulders, or overhangs at depths of 30 m or more. Reported to 200 m and depth may depend on location, habitat or feeding habits. Deep-water individuals may represent a different species. Schools may rise towards the surface in pursuit of zooplankton - their main diet. Identified by the pink to orange colour, but may look blue underwater. Males differ from females in having long and broad ventral fins, usually ending in yellow or bright red, but variable with locations and sometimes ventral fins are all yellow. Adults usually 15–18 cm in length, males largest.

A

A. anthias. Male. Cataluñã, Spain. D. 25 m. L. 14 cm. Patrick Louisy.

B

C

A. anthias. **B** male. Cataluñã, Spain. D. 22 m. L. 10 cm. **C** females. Aquarium, Mediterranean. L. 12 & 15 cm. Patrick Louisy.

Cape Verde Basslet
Anthias cf *anthias*

Undetermined species.

Eastern Atlantic. A deep water species, known from few specimens from Marseillle to Cape Verde. Was thought to be a variation of *Anthias anthias*. Its 3rd dorsal spine is not greatly elongated. Has large eye and caudal fin is forked, its lobes broadly rounded with filaments at tips, elongating with age. Ventral fin long and pointed. Dorsal and anal fins are pointed posteriorly. Colour yellow with pink markings. Length to about 28 cm.

A. cf *anthias.* Male? Length 27 cm. Cape Verde. Cambrain Duarte.

Yellowfin Bass *Anthias nicholsi*

Anthias nicholsi Firth, 1933. Off Virginia, USA.

Western Atlantic, Virginia, USA, Carribean and to northern Brazil. Deep water over rocky substrates. Yellow-orange above and silvery below, head with yellow band from tip of snout to axil and one just above from behind eye. Fins all yellow. Reaches a length of ~25 cm.

A. nicholsi. Length 19 cm (FL). Ron Flescher (NOAA).

Threadnose Basslet *Anthias tenuis*

Anthias tenuis Nichols, 1920. Bermuda.

Western Atlantic, from Bermuda and North Carolina to Gulf of Mexico, Colombia and Venezuela. Deep water reefs at depths of 75 m or more. Sometimes collected for the aquarium trade. Yellow to pale orange above and pinkish below, and iris bright yellow. Length to ~11 cm.

A. tenuis. Aquarium. Female. Length about 8 cm. Scott Michael.

Subfamily Anthiinae, Genus *Holanthias* Günther, 1868

Masculine. Type species: *Anthias fronticinctus* Günther, 1868. An Atlantic genus with 2 species, and a possible new species in Brazil. Body elongated and caudal fin truncate to rounded. Eye relatively small when adult.

1 *Holanthias duplicidentatus* (Miranda-Ribeiro, 1903)	Brazil
2 *Holanthias fronticinctus* Günther, 1868	St. Helena I.

St Helena Seaperch
Holanthias fronticinctus

Holanthias fronticinctus Günther, 1868. St. Helena.

Appears to be restricted to the St Helena Islands. Known depth range is 20–110 m and like most planktivores in this region, it probably schools above rocky reefs. Female plain orange with yellowish fins. Males have filamentous rays on anterior ventral, anal, and central caudal fins. Unlike most other anthiids, the caudal fin is distinctly rounded. Length to 22 cm.

H. fronticinctus. St. Helena Islands. Length 20 cm. Alasdair Edwards.

Brazilian Seaperch
Holanthias duplicidentatus

Anthias duplicidentatus Miranda-Ribeiro, 1903. Rio de Janeiro, Brazil.

Eastern Brazil. Shallow rocky reefs from 15 to at least 40 m depth. Pinkish with irregular broad reddish band below spinous part of dorsal fin, and pale fins. Eye smallish, its diameter less than snout length. Length to 16 cm.

Remarks: previously treated as a synonym of *Pronotogrammus martinicensis*, but this needs further investigation.

H. duplicidentatus. Rio de Janeiro, Brazil. D. 15 m. L. 14 cm. Alfredo Carvalho.

Subfamily Anthiinae, Genus *Hemanthias* Steindachner, 1875 (as subgenus of *Anthias*)

Masculine. Type species: *Anthias (Hemanthias) peruanus* Steindachner 1875. A small genus with 5 species, divided between the Western Atlantic and Eastern Pacific. Generally deep water, mostly around sea-mounts or rock ridges that are swept by strong currents, where they form schools and feed on zooplankton. Most live in depths over 100 m, ranging to about 800 m. Juveniles occasionally shallower. Can be kept in captivity like other planktivores, but they require cool and subdued lighting conditions. Similar to *Anthias*, but lack scales on top of head.

1 *Hemanthias aureorubens* (Longley, 1935)	Florida, Western Atlantic.
2 *Hemanthias leptus* (Ginsburg, 1952)	Alabama, Western Atlantic.
3 *Hemanthias peruanus* Steindachner, 1875	Peru, Eastern Pacific.
4 *Hemanthias signifer* (Garman, 1899)	Panama, Eastern Pacific.
5 *Hemanthias vivanus* (Jordan & Swain, 1885)	Florida, Western Atlantic.

Long-rayed Bass *Hemanthias leptus*

Anthiasicus leptus Ginsburg, 1899. Alabama, USA.

Western Atlantic, south Carolina, USA, to Caribbean waters and probably reaches northern Brazil. Depth range 60–300 m. Adults dusky-yellow with silvery sides and yellow fins. Long filaments on 3rd dorsal-fin spine and ventral fins. Caudal fin slightly forked. A large commercial species, length to 50 cm.

H. leptus. Key West, Florida. Pat O'Donnel (FDEP).

Damsel Threadfin-bass *Hemanthias signifer*

Centristhmus signifer Garman, 1899. Panama, Pacific.

Eastern Pacific, mainland waters, ranging from Baja California to Peru. Mainly in deep water, known range 25–300 m, juveniles occasionally shallower. Adults mainly red with broad yellow outer margins on fins and a very long filamentous 3rd dorsal spine. Caudal fin outer-rays longest. A large species, length to 42 cm.

H. signifer. Salinas, Equador. L. 22 cm. Jimenez Prado.

Red Barbier *Hemanthias vivanus*

Anthias vivanus Jordan & Swain, 1885.Off Florida, U.S.A.

Western Atlantic, from North Carolina to northern Brazil. Deep water, occurring at ~45–600 m. Adults are orange-red and males are marked with yellow spots and violet on sides. Juveniles pale, pinkish with yellow. Long filaments develop on dorsal spines and on corners of other fins with growth. Length to 25 cm.

H. vivanus. Length 13 cm (F.L.). Don Flescher (NOAA).

Split-tail Threadfin-bass *Hemanthias peruanus*

Anthias (Hemanthias) peruanus Steindachner, 1875. Off Peru.

Eastern Pacific, mainland waters, ranging from Baja California to Peru. Occurs around rocky reefs to over 100 m depth. Adults mainly red with broad yellow outer margins on fins and a very long filamentous 3rd dorsal spine. Caudal fin middle-rays longest. A large species, length to 45 cm.

H. peruanus. Panama, Pacific. Ross Robertson.

Streamer Bass *Hemanthias aureorubens*

Pronotogrammus aureorubens Longley, 1935. Florida, USA.

Western Atlantic, Florida, USA, Caribbean and to northern South America. Occurs over rocky bottom, depth range 100–600 m. Reddish with silvery sides. Fins pale pinkish. Dorsal fin without long filaments and notched in centre over posterior spines. Caudal fin large and deeply lunate, outer-rays very long. Length to about 30 cm.

H. aureorubens. Length 24.2 cm (F.L.). Don Flescher (NOAA).

SUBFAMILY ANTHIINAE, GENUS *Pronotogrammus* Gill, 1863

Masculine. Type species: *Pronotogrammus multifasciatus* Gill, 1863. A small genus with 3 deep-dwelling species, one in Atlantic and two in east Pacific. Body elongate and caudal fin truncate to strongly lunate.

1 *Pronotogrammus eos* Gilbert, 1890 — Gulf of California, Mexico

2 *Pronotogrammus martinicensis* (Guichenot, 1868) — West Indies

3 *Pronotogrammus multifasciatus* Gill, 1863 — Baha California, Mexico

Rough-tongue Seaperch *Pronotogrammus martinicensis*

Aylopon martinicensis Guichenot, 1868.

Western Atlantis, ranging from Bermuda and North Carolina to northern Brazil. Deep rocky reefs, known from a depth range of 60 to 600 m. Adult pink with yellow markings. Caudal fin truncate to slightly forked with filamentous tips. Length to 20 cm.

P. martinicensis. 15 cm. Ron Flescher (NOAA).

P. martinicensis. Aquarium from off Florida. Scott Michael.

Faintly-barred Barbier *Pronotogrammus multifasciatus*

Pronotogrammus multifasciatus Gill, 1863. Baha Cal., Mexico.

E. Pacific, ranging in mainland waters from southern California to northern Peru. Occurs over deep rocky bottom in a depth range of 35 to 200 m. Orange above and purplish to silvery below, with dark reticulations on upper sides sometimes forming pattern of vertical bars. Caudal fin lunate to moderately forked with rounded corners. Ventral fins with long filamented tip. Length to 26 cm.

P. multifasciatus. Salinas, Equador. L. 16 cm. Jimenez Prado.

Lunare-tail Barbier *Pronotogrammus eos*

Pronotogrammus eos Gilbert, 1890. Gulf of California, Mexico.

Eastern Pacific, mainland waters, southern Gulf of California to Panama. Occurs over deep rocky bottom in a depth range of 100 to 300 m. Orange above and silvery below, with yellow vertical streaks over scale-rows and yellow spots on fins. Caudal fin strongly lunate. Length to about 20 cm.

P. eos. Adult. Panama, Pacific. Ross Robertson.

Pseudanthias hypselosoma. Spawning. Anthiids are pelagic spawners, releasing gametes in open water. Males court females near the substrate and when responded to, the pair swim up towards the surface to spawn. Larval stages are often prolonged until reefs are reached. Carried by oceanic currents, some species may expatriate far from their place of birth. Photos: Tomonori Hirata, Japan.

SUBFAMILY ANTHIINAE, GENUS *Pseudanthias* Bleeker, 1871

Masculine. Type species: *Anthias pleurotaenia* Bleeker, 1857. Comprises about 70 species, several of which are undescribed. Some species have separate populations that may represent different species and no doubt more new species will be discovered in the near future. As presently defined, the genus contains the most spectacular and largest group of basslets. The majority, the fairy basslets, are placed in the genus *Pseudanthias*. However, there are a number of distinctive groups that were treated by the earlier ichthyologists as different genera, but now treated as subgenera. It made a lot more sense having different genera for the different groups. As, after all, taxonomy is a human invented system that is often in conflict with itself. Species may simply show differences in colour only, whilst biologically they are clearly different. New techniques coming on-line, such as using DNA, are likely to reverse the present "lumping together by workers who can not see the difference. Although Bleeker described the genus a long time ago, until recently most members were described as *Anthias*. Now, the genus *Anthias* is regarded as endemic to the Atlantic region and *Pseudanthias* as being Indo-Pacific. The use of 'anthias' for these basslets as a popular name is also recent and should be reserved as a scientific name only. The species of *Pseudanthias* have been published for many years as basslets, fairy basslets, seaperches (Australia) and goldies (South Africa) and I see no reason to change this.

n.i. = not included

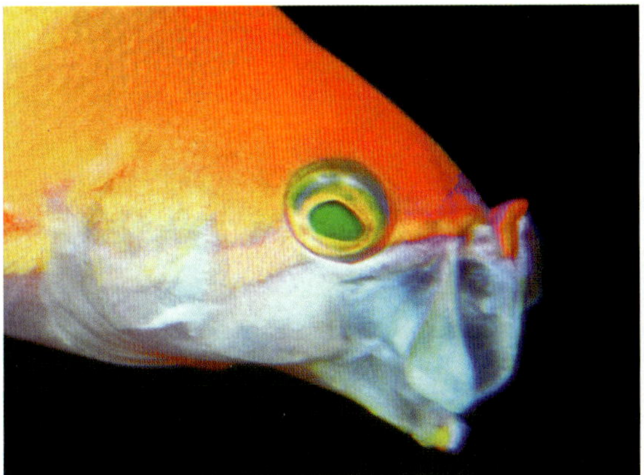

Pseudanthias (Mirolabrichthys) dispar. Anthiids exercise their mouth by yawning to keep their muscles flexible and ready to strike fast, often in quick succession, at prey when feeding on zooplankton.

Pseudanthias (Mirolabrichthys) tuka. Male with forward projecting tip on the snout. A peculiar male-feature that is typical for this subgenus, but in some species this characteristic is less obvious than in others.

SUBGENERA

Amongst the various forms or subgenera, the species of the subgenus *Mirolabrichthys* are the most distinct in having a fleshy, forward-projecting tip on the snout, most obvious in males (see picture right-centre). These are mainly found in pristine oceanic locations, usually along outer reef walls. Most species in this subgenus are difficult to keep successfully in captivity. They require top water-quality and careful selection of companion species. However, this is also the case for the more tolerant *Pseudanthias* that are generally easier to keep, but all need a regime of constant feeding of small prey. In the wild they feed most of the day, swimming above the bottom in pursuit of zooplankton. They have a small but protractile mouth to provide a strong suction to catch prey. When watching these fishes feed, it is seen as a quick flick of the mouth. When retreating to reefs, the mouth is exercised regularly by yawning (see picture above). The species of the *P. squamipinnis*-complex have scaly fins, as the species name suggests, and together with *P. hutchtii* and *P. mortoni*, are placed in the subgenus *Franzia.* In addition to scaly fins, their caudal fin is lunate and the 3rd dorsal fin spine is often very long, but the fin variation can be observed in other genera to various extents. The remaining *Pseudanthias* are in the subgenus of the same name.

Pseudanthias bicolor. Mouth-to-mouth fights are occasionally observed in basslets or other serranids. They have small teeth in their jaws and speed and strength determines the victor. Photo: Hawaiian Is. by John Hoover.

IN THE AQUARIUM

To be successful in keeping fairy basslets, it would help to know from where and from what habitat specimens originated. Coastal species tend to be more tolerant to fluctuating conditions and this

Mixed Basslets schooling into current and feeding on zooplankton: *Pseudanthias cheirospilos, dispar, huchtii*, and *hypselosoma*. Flores, Indonesia. Coastal reef crest, depth about 6 m. On the Malay Peninsula, the anthiids are well represented and numerous species share the same reefs, showing no aggression to each other when feeding. Numbers mean safety, especially when rising above the bottom to feed and form a possible target for a roving predator. Unless one makes a mistake, the many eyes and other sensors give predators little chance.

even goes for individuals of the same species. Some species are widespread, but have geographical variations and in some cases similar species may be treated under the same scientific name. For example, the name *Pseudanthias squamipinnis* is commonly applied to 4 species divided over certain regions of the Indo-Pacific and the true species is restricted to the Red Sea and Indian Ocean. It can be distinguished from its siblings in having a yellow posterior margin on the caudal fin. Specimens offered for sale can originate from many different places that have a large range with regards to temperature and water properties. In some areas individuals may tolerate temperatures as low as 15°C during the winter months and summer temperatures remain in the low 20's during the warmest periods. Such individuals will struggle to survive in the normal tropical coral-reef aquarium. Therefore it is important to know where specimens come from, and this goes for most fish. Knowing temperature regime and water conditions would help in providing the best condition in captivity. For example, specimens of the same species from the Maldives are probably conditioned to higher levels of oxygen and better quality of water, than to those from Sri Lanka where coastal regions are influenced by large rivers and visibility, salinity etc change with the seasons. The same goes for other species that are widespread and different populations are regarded as not being different enough to warrant different species or subspecies status, usually showing only minor colour differences. However, when individuals are unusually large for their species they are likely to originate from cooler than normal habitats.

Species in the subgenus *Mirolabrichthys* are usually the most difficult to keep, firstly because they need very good water quality, and secondly because of their food requirements. These fairy basslets live mostly on outer reefs where currents carry plankton at all times and feeding goes on as long as there is light. These conditions are impossible to provide in captivity unless the keeper has lots of live food available and works out a way to make it available at all times. I have noticed that when food is offered on a few times a day basis, many individuals will over-eat, showing an extended gut. As this does not happen when food is there all the time, and this surely must have some effect on

Orange Basslets *Pseudanthias* cf *cheirospilos*, schooling and feeding on zooplankton that drifts over the reef in the current. In oceanic locations the diversity of species is much lower compared to mainland regions. Usually a single basslet-species forms enormous schools in areas of the Pacific, such as Fiji, in the Indian Ocean, the Maldives, and in the Red Sea where its sibling, the Lyre-tail Basslet *Pseudanthias squamipinnis,* is probably the champion in forming numbers (see p. 25). Photo: Fiji. Gary Bell.

their health. Ideally food is cultured, such as mysids, amphipods, or some other small crustacean. Of course this requires space and it becomes a full-time job. Adding the food to the aquarium is also something that needs to be thought about. Perhaps using a partition containing small holes that the food can swim out of, but the fish can not get through. I used to siphon water from the aquarium into a container and put the food in that, then empty this in to the back of the tank behind some rocks. This enabled the food to settle down and become more of a challenge to be captured. Those people living on the coast may find ways of catching live food and mysids do occur just about everywhere. There are many mysid species, some of which swim on sand patches and some in caves. The cave dwellers are quick swimmers and the best to feed fairy basslets on as they are difficult to catch. It should be kept in mind that cold-water mysids do not live long in tropical conditions. In such cases small amounts should be given on a regular basis or if large amounts can be caught (often seasonally), the surplus can be frozen. The diet should be varied, especially when culturing food such as mysids. These should be enriched by different substances to vary the diet, and additional foods should be supplemented. Crab or other crustacea meats, scallops and mussels can be kept frozen and chopped finely, and other local sea-foods could be utilised.

Most species do well in small groups, but space restrictions may cause friction between adults. The larger the aquarium the better for keeping small groups of a certain species or having mixed fairy basslets. They require lots of hiding places in which they can take refuge, but also plenty of room to swim openly. Without enough hiding places and a lack of swimming space, most species become aggressive. Ideally they are kept in a large 'living-reef' aquarium with other fishes of similar size, such as the more cryptic rock-basslets, pipefishes, or other planktivores. Different species of fairy basslets can be kept together, but again, this requires lots of space. In the wild they co-occur without any obvious problems and seem to share close spaces and feed together with only the occasional display of large fins. Introducing new individuals to an established aquarium that already contains basslets needs to be done with care, especially with large adults.

Indian Lyre-tail Basslet *Pseudanthias squamipinnis*

Serranus (Anthias) squamipinnis Peters, 1855. Mozambique.
Anthias gibbosus Klunzinger, 1884. Red Sea.

Widespread Indian Ocean and Red Sea. Various rocky and coral reef habitats. In some areas, such as in the Red Sea, they occur in massive numbers when feeding on zooplankton (see photo on opposite page by Roger Steene). Adults with long filamentous third dorsal spine and caudal fin deeply lunate with lobes increasingly elongating with age. Females are bright yellow to deep orange with yellow fins. Males purplish red with pale sides, most scales with a yellow central spot. The caudal fin has a yellow posterior margin that is most obvious in large individuals which readily distinguishes it from its Pacific siblings. Length to 12 cm, excluding filamentous tips. It grows largest in the southern limit of its range off the eastern coast of South Africa where it is called Sea Goldie.

A
P. squamipinnis. Aliwal Shoal, South Africa. D. 20 m. L. 12 cm.

B
P. squamipinnis. Maldives. Depth 20 m. Length 11 cm.

C
P. squamipinnis. Maldives. D. 20 m. L. 10 cm.

D
P. squamipinnis. Red Sea. Depth 20 m. Length 11 cm.

E
P. squamipinnis. Maldives. D. 20 m. L. 9 cm.

F
P. squamipinnis. Red Sea. Depth 20 m. Length 10 cm.

G
P. squamipinnis. Maldives. D. 20 m. L. 10 cm.

Indonesian Lyre-tail Basslet
Pseudanthias cheirospilos

Anthias cheirospilos Bleeker, 1857. Ambon.

Widespread Indonesia and Philippines. It occurs commonly on shallow coastal reefs and outer reef lagoons, where it schools in current prone areas. Juveniles and females mostly orange with yellow ventral fins and yellow lobes on caudal fin. Males reddish or purplish, body scales with yellow spot in centre, anal fin bluish with yellow spots and caudal fin with a thin, blue posterior margin. Length to 10 cm.

Remarks: previously included with *Pseudanthias squamipinnis*. Differs in colour and morphology. The more obvious differences are the blue, versus yellow posterior margin on the caudal fins and the lateral-line possessing many auxiliary scales, versus few, as well as having an almost truncate caudal fin, versus deeply lunate.

A

P. cheirospilos. Flores, Indonesia. Depth 9 m. Male, length 10 cm.

B

C

P. cheirospilos. Bali, Indonesia. Depth 4 m. Juveniles, lengths 4–5 cm.

D

P. cheirospilos. Bali, Indonesia. Depth 4 m. Schooling with *P. hypselosoma*.

P. cheirospilos. Bali, Indonesia. Depth 8 m. Length about 10 cm.

P. cheirospilos. Indonesia. **F** & **G** Flores. Males. Length 10 cm. **H** Bali. Depth 12 m. Females, length 8 cm.

Orange Basslet *Pseudanthias* cf *cheirospilos*

Coral Sea, and New Guinea to central Pacific, ranging to subtropical zones of eastern Australia. A form of *Pseudanthias cheirospilos* that appears to represent an undescribed species. Rocky and coral reefs, adults usually in 10 m depth or more, sometimes in massive schools. Nuptial males are deep purple in colour. Small males have a yellow anal fin and the entire caudal fin margin is electric blue. Juveniles and females are bright orange, the fins marginally paler or yellowish. Length to 10 cm.

Remarks: previously confused with *Pseudanthias squamipinnis*, but differs considerably in colour, e.g., males lack yellow spots on scales, and their caudal fin is not deeply lunate.

A *P.* cf *cheirospilos*. Milne Bay, PNG. Depth 10 m. Length 10 cm.

B *P.* cf *cheirospilos*. NSW, Australia. D. 20 m. Females.

C *P.* cf *cheirospilos*. Milne Bay, PNG. Depth 10 m. Length 10 cm.

D *P.* cf *cheirospilos*. Qld, Australia. D. 20 m. Male and females.

E

F

P. cf *cheirospilos*. Montague I., NSW, Australia. Depth 12–16 m. **E** male. Length 10 cm. **F** female, 6 cm.

Japanese Lyre-tail Basslet *Pseudanthias nobilis*

Anthias nobilis Franz, 1910. Misaki, Japan.

Southern Japan, ranging north into subtropical zones and to the south probably ranging to Taiwan. The geographical range uncertain as previously confused with *Pseudanthias squamipinnis.* Found on rocky reefs, adults at moderate depths, usually over 10 m, and juveniles often in shallow protected bays. This species has a large caudal fin that becomes deeply lunate with age and large, elongated ventral fins. Males are densely spotted with yellow over the body and posteriorly extending onto fins. In nuptial males the anterior part of the dorsal fin looks white when observed underwater. The largest of the '*squamipinnis*-complex', reaching at least 14 cm, excluding caudal filaments.

A

P. nobilis. Nuptial male. Izu, Japan. Depth 20 m. Length 14 cm.

B

C

P. nobilis. Juveniles. Kochi Pref, Japan. Tomonori Hirata.

D

E

P. nobilis. Japan. **D** Kochi Pref. Schooling. Depth 8 m. Tomonori Hirata. **E** Izu, nuptial male. Depth 15 m. Length 14 cm.

F

G

P. nobilis. Hachijio Is, Japan. **F** small colourful male with females. According to photographer it participates in spawning, but is chased away by large males when discovered. Tomoyo Mizutani. **G** induvidual in process of changing sex. D. 12 m. L. 9 cm. Shouichi Kato.

A

P. huchtii. Flores, Indonesia. Depth 6 m. Length 11 cm.

Green Pacific Basslet *Pseudanthias huchtii*

Pseudanthias huchtii Bleeker, 1857. Ambon.

Widespread tropical West Pacific, but replaced by similar species in Australia and Papua New Guinea. Coastal coral reefs and lagoons on outer reefs. Usually schooling on reef crests in 3–20 m depth. Juveniles and females mostly yellow, grading to a dusky colour dorsally and eye distinctly yellow. Males grey-green with a bright red band from eye to pectoral fin base and extending onto fin and ventral fins with a red and broad leading edge. Median fins submarginally red and dorsal filament red. Females mostly yellow. Length to 11 cm.

B

C

P. huchtii. Indonesia. **B** Flores. Male with a double filament, a rare feature. **C** Bali. Female.

A

P. mortoni. Milne Bay, PNG. Depth 7 m. Nuptial male. Length 9 cm.

Pink Pacific Basslet *Pseudanthias mortoni*

Anthias mortoni Macleay, 1883. New Guinea.

Eastern Australia to Papua New Guinea and Solomon Islands. Coastal coral reefs and lagoons on outer reefs. Usually schooling in coral-rich habitats at 6–25 m depth. Juveniles and females mostly pale yellow, grading to grey dorsally and eye distinctly yellow. Female has yellow band from eye to pectoral fin base and on outer margins of anal and caudal fins. Males yellow-pink to purple on sides and a red to purple band from eye to pectoral fin and on leading edge of ventral fins. Median fins mostly yellow, grading to dusky submarginally and pale blue outer margins on caudal fin. Females pale yellow with bright-yellow band from eye to pectoral fin base and yellow margins on median fins. Male's dorsal filament dusky to pale brown. Length to 11 cm.

B

C

P. mortoni. **B** Solomon Is. Depth 15 m. Length 10 cm. Roger Steene. **C** Great Barrier Reef, Australia. L. 85 mm. Phil Woodhead.

Aurora Basslet *Pseudanthias calloura*

Pseudanthias calloura Ida & Sakaue, 2001. Palau.

Primarily known from Palau, but recently reported from northern Sulawesi, Indonesia. A deep water species found on rocky drop-offs in depths of 55 m or more where in small groups. A spectacular species with boldly marked colours of blue, yellow and orange, males with white band from below eye to end of anal fin. Length to about 10 cm.

Remarks: also known as Harlequin Basslet, but this name is already in use for *Pseudanthias connelli*.

A

B

C

P. calloura. **A** & **B** Palau. Depth 55–65 m. Lengths 9 cm. Hiroshi Nagano. **C** Palau. Ryoichi Satoh.

Harlequin Basslet *Pseudanthias connelli*

Anthias connelli Heemstra & Randall, 1986. Durban.

Apparently restricted to KwaZulu-Natal, South Africa. Seems to favour remote reefs in deep water and mainly observed on shipwrecks in the Durban region. Usually mixes with *Pseudanthias squamipinnis* when feeding on zooplankton, often rising high above the substrate when currents are running. Females are difficult to spot amongst females of *P. squamipinnis*, but have a different caudal fin with short corners and a red posterior margin. Males are very distinct in colouration with their broad double dark band on the head that continues onto the body, ending abruptly at vertical above anus, especially during displaying. Length to 12 cm.

A

P. connelli. Aliwal Shoal, South Africa. Depth 25 m. Nuptial male, ~12 cm.

B

C

P. connelli. Aliwal Shoal, South Africa. Depth 25 m. **B** female. Length ~11 cm. **C** male. Length~12 cm.

A

P. townsendi. Fully coloured male. Northern Oman. Depth ~20 m. John Hoover.

Oman Basslet *Pseudanthias townsendi*

Anthias townsendi Boulenger, 1897.
Mekran coast, Iran.

Arabian Gulf, Gulf of Oman and Arabian Sea. Rocky reefs with rich invertebrate growth, usually in depths of about 20 m or more, rising to shallower depths to feed on zooplankton. Males are easily identified by their longitudinal bands and tail pattern, but females are similar to several other species. Since males normally swim or mix with groups of females, they can be identified by association. Length to 11 cm.

B

C

P. townsendi. **B** male. Northern Oman. **C** females. Southern Oman. Phil Woodhead.

D

P. townsendi. Young male with female behind (*Apogon fleurieu* below). Northern Oman. Phil Woodhead.

Red-Sea Basslet *Pseudanthias taeniatus*

Anthias (Pseudanthias) taeniatus Klunzinger, 1884. Egypt, Red Sea.

Endemic to the Red Sea. Occurs on reef slopes rich with corals, often forming large aggregations in which females greatly outnumber males. Males are readily identified by the banded pattern and their mostly white caudal fin, but females are more plain, shading from orange above to pink below, and have red tips on the caudal fin. Length to 13 cm.

A *P. taeniatus*. Male. Egypt, Red Sea. Depth about 20 m. Length 11 cm.

B

C

P. taeniatus. Females. Egypt, Red Sea. Depth about 20 m. Length 9–10 cm.

D

E

F

P. taeniatus. Red Sea. **D** Aqaba. Changing sex. Thomas Paulus. **E** & **F** Egypt. Males. Depth about 20 m. Lengths about 12 cm.

Marcia's Basslet *Pseudanthias marcia*

Pseudanthias marcia Randall & Hoover, 1993. Oman.

Known only from the Oman region, but it may occur in the Maldives (see next page). Abundant on the south-western coast, in particular on a drop-off near Rahah Bay where it occurs in schools. Males are unusual in having a prolonged and filamentous upper lobe on the caudal fin that is not mirrored by the lower lobe. Juveniles and females are similar to the sympatric *Pseudanthias townsendi* and can be distinguished from it by its uniformly pink colour, versus a more yellowish posterior part of the body in *P. townsendi*. Length to about 12 cm.

A

P. marcia. Male. South-western Oman. Depth ~20 m. John Hoover.

B

C

P. marcia. Male. South-western Oman. Depth ~20 m. John Hoover.

D

P. marcia. Juveniles. South-western Oman. Depth ~20 m. John Hoover.

E

P. marcia. Males and female upper-left. South-western Oman. Depth ~20 m. John Hoover.

Reticulate Basslet *Pseudanthias* sp 1

Undetermined species.

According to Scott Michael and Hiroyuki Tanaka, the fish received in the USA and Japan were thought to be shipped from the Maldives (based on information supplied by the importers). The specimen photographed in Japan looks virtually identical to *Pseudanthias marcia* and this suggests that it may just be a colour variation, being more colourful. If indeed from the Maldives, this species may compete with several congeners and this may create a social pressure between them, promoting brighter colours in males.

A

P. cf *marcia*. Male. Aquarium. From Maldives? Hiroyuki Tanaka.

B

C

P. cf *marcia*. Males. Aquarium. From Maldives? Scott Michael.

Bloodspot Basslet *Pseudanthias* sp 2

Undetermined species.

Izu Islands, Pacific coast of southern Japan, to the Philippines, and may occur in northern Indonesia, but latter reports could be based on other similar species with a red-blotch on the dorsal fin. Depth range and habitat not well known, but adults probably occur fairly deep (over 50 m). The way the red spot on the dorsal fin base is equally divided between the fin and back distinguishes it from other species with a red spot on the dorsal fin (usually at ~5th–9th spine). Length to about 10 cm.

A

P. sp 2. Kumejima, Japan. Depth 63 m. Length 10 cm. Yasuaki Miyamoto.

B

C

P. sp 2. Oshima, Japan. Depth 42 m. Length 25 mm. Yasuaki Miyamoto. **C** aquarium. Philippines? Scott Michael.

Pale-bar Basslet *Pseudanthias engelhardi*

Anthias engelhardi Allen & Starck, 1982.
Escape Reef, Great Barrier Reef, Australia.
Pseudanthias carlsoni Randall & Pyle, 2001. Fiji.

Escape Reef to Milne Bay, Papua New Guinea and east to Fiji. Deep reef habitats with rich coral and sponge growth at depths of 40 m or more. Usually in small groups of mixed sex, but females greatly outnumber males. Adults pinkish with or without a red spot midway on the spinous section of the dorsal fin, that looks black in deep water, and a pale vertical red bar on side above abdomen. The red bar in males may turn white in nuptial males. Length to just over 10 cm.

Remarks: very similar *Pseudanthias* cf *engelhardi* from off the western coast of Australia (opposite page) may also represent this species, primarily being different in colour: not showing any sign of a vertical bar on its sides and a dorsal-fin spot that appears mainly during display.

A

P. engelhardi. Holotype (female?). Length 102 mm. Roger Steene.

B

C

D

E

P. engelhardi. **B–E** Milne Bay, PNG. Depth 50 m. Lengths about 9–10 cm.

F

G

P. engelhardi (*carlsoni* form). Fiji, Depth 35 m. **F** nuptial male**. G** female. Jerry Allen.

Shoals Basslet *Pseudanthias* cf *engelhardi*

?Pseudanthias engelhardi.

Bernier I. to Rowley Shoals, off north-western Australia. A common species of deep slopes adjacent to outer-reef walls. Females form large aggregations and remain close to the substrate to feed on plankton. Males in vicinity of groups of females. Only males intermittently show a red spot in the dorsal fin. Length to 9 cm.

Remarks: appears to be closely related to, and perhaps is only a geographical variation of, *Pseudanthias engelhardi.*

A

B

C

D

P. cf *engelhardi.* Rowley Shoals, WA, Australia. Depth 54 m. Length 60–75 mm, with and without dorsal-fin spot.

E

P. cf *engelhardi.* Rowley Shoals, WA, Australia. Depth 54 m. Length 40–50 mm.

A

P. luzonensis. Flores, Indonesia. Depth 30 m. Male, length 14 cm.

Luzon Basslet *Pseudanthias luzonensis*

Anthias (Pseudanthias) luzonensis Katayama & Masuda, 1983. Philippines.

West to Central Pacific, southern Japan to northern Australia and east to Fiji. Inshore reefs on rich coral slopes, usually at 20+ m depth. Occurs solitary or in loose small groups of mixed sexes. Males pink, darkest on top of head, with 3 or 4 yellow mid-lateral lines that are often irregular. They have a large red blotch on the dorsal fin from behind the 7th spine, sometimes reaching as far as the soft rays, and red blotch on posterior edge of caudal fin. These blotches look black when seen in natural light when in deep water. Females are plain pink with a darker pink spot on each scale, darkest dorsally, and a yellow line from eye to base of pectoral fin. Males usually larger than females. Length to about 14 cm.

B

C

P. luzonensis. Indonesia. **B** south of Bitung, Sulawesi. Depth 45 m. Male, length 14 cm. **C** Bali. D. 25 m. L. 65 mm.

D

E

P. luzonensis. **D** Fiji. Depth 32 m. Male, length 14 cm. Jerry Allen. **E** Bali, Indonesia. Female, length 12 cm.

Manado Basslet *Pseudanthias manadensis*

Anthias manadensis Bleeker, 1856. Manado, Sulawesi.

Known only from the single type specimen, not yet linked with any living form. Bleeker's meristics are D X,17 (–18, probably referring to last 2 rays, sharing a common joint); P 2/17; A III, 7–8; longitudinal scale rows 45. Total length 106 mm. His illustrations, produced by various artists, are reputed for their accuracy and it is probably only a matter of time until this species is found. Amazingly, Bleeker obtained species that are only now being re-discovered. This distinctive species is still elusive. However, *Pseudanthias venator* (Snyder, 1911) from Japan may be the male phase, thus a junior synonym, but is also known only from a single specimen.

P. manadensis. After Bleeker.

Pink Basslet *Pseudanthias hypselosoma*

Pseudanthias hypselosoma Bleeker, 1877. New Guinea.

Widespread Indo-West Pacific, expatriating to cool subtropical zones during warm currents. Coastal to outer reefs in depths from ~6 m down to moderate depths. Often forms very large schools above remote coral-heads in current-prone habitats when feeding on zoo-plankton. Mostly plain pink, but males display with intense colours (**D**). Juveniles and females have a red posterior margin on the caudal fin with red tips, and males feature a red blotch on the spinous part of the dorsal fin. Length to 12 cm.

Remarks: nuptial males of the Japanese form have a yellow streak on belly below pectoral fin base and red bands in dorsal and anal fins (**E**). This form was described as *Anthias (Pseudanthias) truncatus* by Katayama & Masuda, 1983, from Kerama.

A

P. hypselosoma. Male. Tulamben, Bali. Depth 7 m. Length 11 cm.

B

C

P. hypselosoma. Nuptial male. Great Barrier Reef, Australia. Phil Woodhead.

D

P. hypselosoma. NSW, Australia. Juvs., 35–45 mm.

E

P. hypselosoma. About to spawn. Kochi Pref, Japan. Tomonori Hirata.

F

G

P. hypselosoma. **F** PNG. Feeding on plankton during tidal currents. **G** Bali, Indonesia. Nuptial males. D. 7 m. L. 10 cm.

Heemstra's Basslet *Pseudanthias heemstrai*

Pseudanthias heemstrai Schuhmacher, Krupp & Randall, 1989. Aqaba, Red Sea.

Endemic to Red Sea, found along fringing reefs in depths over 20 m. Often in very large aggregations of similar sized individuals over its entire size range, even from the smallest juveniles. Juveniles are bright yellow above and deep pink below the lateral-line, whilst the dorsal and caudal fins are yellow. Females more yellow, grading into pink below lateral-line and fins more pointed. Males very distinct with a mostly yellow-orange head and a red caudal fin with a yellow posterior margin and blue tips. Nuptial males display a white lateral stripe and caudal fin turns bright red. Length to 13 cm.

P. heemstrai. Males. **B** nuptial. **C** including a *P. fasciatus*. Aqaba, Red Sea. Depth 30–50 m. Thomas Paulus.

P. heemstrai. **D** females with two males above. **E** juveniles. Aqaba, Red Sea. Depth 30–50 m. Thomas Paulus.

P. heemstrai. Adults of mixed sex, massing near their shelter. Aqaba, Red Sea. Depth 25 m. Thomas Paulus.

Lilac-tipped Basslet
Pseudanthias rubrizonatus

Anthias rubrizonatus Randall, 1983. Solomon I.

Widespread West Pacific, ranging to sub-tropical zones. Sporadic distribution. Usually in small aggregations in deep current channels with remote coral heads, often mixing with *Pseudanthias hypselosoma* which is much more common in many areas of Indonesia and Papua New Guinea. Juveniles enter harbours and often on silty rocky reefs. Juveniles and females recognised by their bluish median fins and lilac tipped dorsal spines and dusky centres to their scales. Males readily identified by the broad red band on their sides and yellow caudal fin with blue margins, but latter more red in Japan. Length to 12 cm, but 15 cm in Japan.

A

P. rubrizonatus. Sulawesi, Indonesia. D. 30 m. Nuptial male, 12 cm.

B

C

P. rubrizonatus. Flores, Indonesia. Depth 15 m. **B** sucking in plankton, note long gill-rakers. **C** large male, length 12 cm.

D

P. rubrizonatus. Kochi Pref., Japan. D. 45 m. L. 15 cm. T. Hirata.

E

P. rubrizonatus. Sydney, Australia. Juv. 40 mm.

F

G

P. rubrizonatus. Japan. **F** Kochi Pref. Depth 54 m. Female, length 8 cm. **G** Ehime Pref. D. 12 m. Small juvenile. Tomonori Hirata.

A

Yellow-eye Basslet *Pseudanthias lunulatus*

Anthias lunulatus Kotthaus, 1973. Somalia.

Known from western Indian Ocean, African coast to Mauritius, Maldives, and Red Sea. A deep water species, only occasionally seen on reef outcrops at depths as shallow as 50 m. Caudal fin large, deeply lunate in both sexes. Female pinkish, eye distinctly yellow, and a yellow stripe running from tip of snout to pectoral fin base. Males feature a yellow triangular saddle below spinous dorsal fin that extends vertically into a narrow and short band that may reach abdomen. Anal fin elongates with age, reaching to below caudal fin. Length to 12 cm.

B

P. lunulatus. **A** male. Length 12 cm. **B** male and female. Sanganeb Atoll, Sudan, Red Sea. Depth 60–65 m. Thomas Paulus.

C

D

P. lunulatus. **C** female. Maldives. Length 11 cm. Jörg Aebi. **D** juvenile. Sanganeb Atoll, Sudan, Red Sea. Thomas Paulus.

E

F

P. lunulatus. Males. Maldives. Depth 50 m. Lengths 10–11 cm. Jörg Aebi.

Bali Basslet *Pseudanthias* cf *lunulatus*

?*Anthias lunulatus* Kotthaus, 1973. Somalia.

Known only from Bali, Indonesia. Only occasionally seen on reefs at depths as shallow as 40 m, but probably normally occurs much deeper. Juveniles and females pale, pinkish, eye yellow and a yellow stripe from tip of snout to pectoral fin base, and caudal fin large and deeply lunate. Males feature a yellow saddle, extended below with a vertical band below spinous dorsal fin. Length to 14 cm.

Remarks: possibly just a geographical variation of *Pseudanthias lunulatus*, but there are differences in colour, and the dorsal and anal fins are rounded.

A

B

C

P. cf *lunulatus*. Bali, Indonesia. **B** juvenile (parasites on tail) Akira Ogawa. **A** & **C** male. D. 40 m. L. 14 cm. Gary Bell.

Nusa Penida Basslet *Pseudanthias* sp 3

Undetermined species.

A rarely observed deep water species known from Bali, Indonesia. Usually at depths over 40 m, and in strong current habitats. Females plain pink with yellow head, parallel violet lines from tip of snout to pectoral-fin base, and with an orange-red spot below the base of the dorsal fin. The spot elongates and gradually extends downward with age and forms a bar in males. Caudal fin with blue and yellow, rounded tips. Length to about 12 cm.

Remarks: this basslet has been confused with *Pseudanthias engelhardi,* a somewhat similar species from the Coral Sea region.

A

P. sp 3. Bali, Indonesia. Depth 40 m. Length 12 cm. Akira Ogawa.

B

C

P. sp 3. Bali, Indonesia. Depth 40 m. **B** female. **C** male. Length 12 cm. Gary Bell.

Double-bar Basslet *Pseudanthias* sp 4

Undetermined species.

Known only from Palau and Marshall Is, but most likely to be more widespread. A deep water species, usually at depths over 50 m, that appears to be closely related to *Pseudanthias engelhardi*. Juveniles and females seem identical to that species, being plain pinkish with a large yellow eye. Males are distinct in having a double vertical yellow or reddish bar on the side, and intermittently a red spot in the dorsal fin. Length to 11 cm.

A *P.* sp 4. Male. From Marshall Is. Length 10 cm. Brian D. Greene.

B *P.* sp 4. Female. Palau. D. 50 m. L. 9 cm. Hiroshi Nagano.

C

D

P. sp 4. Males. Marshall Is., Arno-Atoll. **C** depth 50 m. **D** depth 68 m. Lengths ~10 cm. Yasuaki Miyamoto

Hutomo's Basslet *Pseudanthias hutomoi*

Anthias hutomoi Allen & Burhanuddin, 1976.
Maluku Islands, Indonesia.

Indonesia. A rare species, known only from trawled type-specimens from 54 m depth, and a few specimens have appeared in the aquarium trade. 3rd dorsal spine longest, soft dorsal rounded with 14 rays, and caudal fin is shallowly forked. Females plain pink-orange with yellowish fins. A red band with pink margins behind eye. Length to about 10 cm.

Remarks: closely related to *Pseudanthias cichlops*.

A

B

C

P. hutomoi. Aquarium. Specimens shipped from Bali, Indonesia. Scott Michael.

Four-spot Basslet *Pseudanthias cichlops*

Serranus cichlops Bleeker, 1853. Sumatra.

Widespread Indonesia, ranging to PNG and Philippines. A deep water species in current-prone areas, forming small aggregations near large coral heads at the base of deep slopes. Usually in depths over 40 m, but some were seen at 20 m in Solor. Third dorsal spine not long and soft-ray part pointed. Juveniles and females with 4 white spots over back. A yellow band behind eye. Length to 12 cm.

Remarks: very similar to *Pseudanthias hutomoi*, but this species is thought to have no more than 14 soft dorsal-rays and a slightly forked caudal fin, versus 15–16 rays and lunate fin, and has a prolonged 3rd dorsal spine.

A

P. cichlops. After Bleeker.

B

C

D

P. cichlops. Indonesia. **C** Bali. D. 60 m. Juvenile, l. 65 mm. Akira Ogawa. **C** & **D** Lembeh Strait, Sulawesi. D. 40 m. Male, l. 12 cm.

E

F

P. cichlops. Lembeh Strait, Sulawesi. Depth 40 m. **E** large juvenile. Length 9 cm. **F** female, length 11 cm.

Silverstreak Basslet *Pseudanthias cooperi*

Anthias cooperi Regan, 1902. Maldives.

Widespread Indian Ocean, ranging east to Sumatra. Adults form small schools at moderate depths. Mostly found on open substrate with low reefs and remote bommies in depths over 25 m. Juveniles and females reddish dorsally to pale pinkish ventrally, a red band broadly along base of dorsal fin, caudal fin mostly red, and ventral fins plain. Males red with yellow anal fin and 'silver' stripe from below eye over cheek. Nuptial males almost white with blood-red caudal and ventral fins, and a red blotch or short bar centrally on sides. Length to 12 cm.

A

P. cooperi. Nuptial male. Maldives. Depth 35 m. Length 9 cm.

B

P. cooperi. Pulau Weh, Sumatra. D. 28 m. L. 10 cm.

C

P. cooperi. Juvenile. Maldives. Depth 35 m. Length 6 cm.

D

E

P. cooperi. Males. **D** Aliwal Shoal, South Africa. Dennis King. **E** Kenya. Depth 18 m. Length 10 cm. Roger Steene.

F

P. cooperi. Females joined by a female *P. hypselosoma* (lower right). Maldives. Depth 35 m. Lengths ~9 cm.

Red Basslet *Pseudanthias kashiwae*

Leptanthias kashiwae Tanaka, 1918. Kochi, Japan.

Widespread Western Pacific. Adults form small groups at depths over 20 m. Mostly found on open substrate with low reefs and remote bommies. Juveniles may occur at shallow depths in coastal reefs. Pink to red, sometimes greenish over back, dorsal fin red with blue margin, and caudal fin reddish with distinct red tips. Ventral fins with red margin on leading edge in larger juveniles and females. Nuptial males variable, pink to red with pale sides, central red bar and caudal fin red. Length to 13 cm.

Remarks: often treated as a synonym of *Pseudanthias cooperi*.

P. kashiwae. Hachijio Is, Japan. D. 35 m. L. 12 cm. Shouichi Kato

P. kashiwae. Spawning. Kochi, Japan. Depth 18 m. Length 8 & 10 cm. Tomonori Hirata.

P. kashiwae. Male variations. Kerama, Japan. Depth 30 m. Length 10 & 12 cm.

P. kashiwae. Juveniles. NSW, Australia. **F** Montague I. (southern-most record). D. 25 m. L. 35 mm. **G** Sydney. L. 45 mm.

P. kashiwae. Flores, Indonesia. Depth 15 m. **H** females. Length 6 cm. **I** male. Length 10 cm.

Hiva Basslet *Pseudanthias hiva*

Pseudanthias hiva Randall & Pyle, 2001. Marquesas.

Known only from the Marquesas Islands. Occurs in rocky habitats at moderate depths, adults usually deeper than 20 m, but juveniles may be much shallower. Juveniles and females plain pink, becoming reddish over the back with growth, and caudal fin with red tips. Males with red caudal fin and very long extensions on tips. Length to 12 cm, excluding filaments.

A

P. hiva. Male. Marquesas. Depth 35 m. John Hoover.

B

P. hiva. Juvenile. Marquesas. Depth 20 m. John Hoover.

C

P. hiva. Male. Marquesas. Depth 35 m. John Hoover.

Yellow-tail Basslet *Pseudanthias olivaceus*

Anthias (Pseudanthias) olivaceus Randall & McCosker, 1982. Rarotonga, Cook Islands.

South Pacific, Cook Island and French Polynesian regions. Outer reef slopes from the shallows to about 35 m depth. Usually in small aggregations near the substrate. Readily identified by the bright yellow caudal fin. Length to 12 cm.

A

P. olivaceus. Rarotonga, Cook Is. Male, length about 11 cm. Malcolm Francis

B

C

P. olivaceus. Niue, South Pacific. **B** female, length ~9 cm. **C** juvenile, length ~45 mm. Malcolm Francis

Red-collar Basslet *Pseudanthias elongatus*

Anthias elongatus Franz, 1910. Yokohama, Japan.

Subtropical southern Japan. Deep coastal rocky reefs in current prone habitats. Forms large schools at depths of 45 m or more. A distinctive species with its long and pointed fins, most of which are extended by filamentous tips. Third dorsal-fin spine long. Females pinkish with numerous yellow spots forming irregular lines on posterior two-third of the body. Nuptial males have a broad red band behind the head and red caudal fin with irridescent blue margins. Juveniles mainly orange. A large species, reaching 16 cm or more, excluding filamentous caudal-fin tips.

A

P. elongatus. Osezaki, Izu Pen., Japan. Depth 45 m. Male, length 15 cm.

B

P. elongatus. Nuptial male. Izu Pen., Japan. D. 45 m.

C

D

P. elongatus. Kochi Pref., Japan. **C** juvenile. L. 35 mm. D. 35 m. **D** large female. L. 13 cm. D. 50 m. Tomonori Hirata.

E

P. elongatus. Osezaki, Izu Pen., Japan. Depth 45 m. Males and females aggregating near the substrate.

Thompson's Basslet *Pseudanthias thompsoni*

Caesioperca thompsoni Fowler, 1923.
Oahu, Hawaiian Islands.

Hawaiian Islands. Usually seen swimming in open water adjacent to reefs to feed on zooplankton, sometimes rising to near surface depths in pursuit of prey. Females are much more numerous than males and occur in small to large aggregations. Females plain pinkish. The multicoloured males grow much larger and develop long caudal filaments similar to the Japanese *Pseudanthias caudalis*. Length to 21 cm, but females about 16 cm.

P. thompsoni. Hawaii. Male, length about 20 cm. John Hoover.

P. thompsoni. Hawaii. **B** male, length ~18 cm. **C** females. Length ~12 cm. John Hoover.

Long-tail Basslet *Pseudanthias caudalis*

Pseudanthias caudalis Kamohara & Katayama, 1959.
Kochi Pref., Japan.

Pacific coast of Honshu and nearby islands to the Ogasawaras. Deep rocky reefs, adults usually at 50 m or more. Rarely noticed species due to plain colouration and usually mixes with other anthiids when feeding above the bottom on zooplankton. Distinguished from other species by its strongly lunate caudal fin, combined with the relatively short ventral fins. Length to 18 cm.

P. caudalis. Izu, Japan. **A** & **B** male. Takashi Hara. **C** female. D. 35 m. Akira Ogawa.

Thread-tail Basslet *Pseudanthias rubrolineatus*

?Anthias georgei Allen, 1976. Off Bernier I., W. Australia.
Anthias rubrolineatus Fourmanoir & Rivaton, 1979. New Caledonia.

Appears to be widespread in the Western Pacific, but with sporadic records, from southern Japan, Indonesia, off Western Australia(?) and Coral Sea, due to deep water habitat preference. Type reported from 400 m depth, but in Japan adults are found as shallow as 50 m. Unusual species in having multiple filaments on caudal fin. Further identified by the red stripe running from just above pectoral fin and crossing the lateral line below soft dorsal fin and ending high up on the caudal peduncle. Length to about 20 cm.

Remarks: appears to be identical to *Anthias georgei*, a possible senior synonym. They share general morphology, including the diagnostic caudal filaments. Live colouration of *A. georgei* is unknown.

A

P. rubrolineatus. Nuptial. Kochi, Japan. D. 55 m. L. 20 cm. Tomonori Hirata.

B

P. rubrolineatus. Oshima, Japan. D. 55 m. L. 9 cm. Yasuaki Miyamoto.

C

P. rubrolineatus. Oshima, Japan. Depth 40 m. Length 3 cm. Yasuaki Miyamoto.

D

E

F

P. rubrolineatus. Kashiwajima, Japan. Depth 55 m. **D** & **E** females ~10–15 cm. **F** nuptial male ~20 cm. Masatoshi Kubo.

A

P. fasciatus. Derawan, Kalimantan. Depth 25 m. Length 16 cm.

Red-stripe Basslet *Pseudanthias fasciatus*

Franzia fasciata Kamohara, 1954. Japan.

Widespread Indo-West Pacific and Red Sea, ranging into sub-tropical zones of Japan, Australia to Fiji, and South Africa. Some populations can tolerate temperatures below 15°C. Rocky reefs, often with large caves or overhangs where swimming upside-down on ceilings, usually in 30–80 m depth. At all stages it can be recognised by the red lateral stripe, but large males may change colour during courtship when the red stripe is absent. Length to 20 cm.

Remarks: most widespread species in the genus, but geographical variations may actually represent a complex of closely related species. South African and Red Sea population most different in colour and females have a straight and thicker red stripe, longer median fins, and a much more lunate caudal fin.

B

C

P. fasciatus. Japan. **B** Osezaki. Depth 50 m. Length 16 cm. **C** Izu Oceanic Park. Depth 25 m. Length 12 cm.

D

E

P. fasciatus. **D** Montague I., NSW. Depth 25 m. Length 35 mm. **E** same individual as adult, grown in aquarium.

P. fasciatus. Males. **F** NSW, Australia. D. 20 m. L. 18 cm. Gary Bell. **G** Fiji, South Pacific. D. 44 m. L. 16 cm. Jerry Allen.

P. fasciatus. **H** grown from juvenile collected at Montague I., length 11 cm. **I** male. Loloata, PNG. D. 25 m. L. 18 cm. Roger Steene.

P. cf *fasciatus*. Red Sea form. **J** females with *P. heemstrai*, **K** female, **L** & **M** males. Aqaba. Depth 55–65 m. All by Thomas Paulus.

Two-spot Basslet
Pseudanthias bimaculatus

Anthias bimaculatus Smith 1955. Mozambique.

Widespread Indian Ocean, ranging east to at least Bali. Coastal to outer reefs along the base of drop-offs at 10 to 60 m depth. Highly variable species, showing different or stronger colour patterns when living together with closely related or similar looking species to enhance recognition by females that swim in mixed-species groups. This may lead to further speciation. Closely related species such as *Pseudanthias pleurotaenia* and *P. bimaculatus* may spawn at the same time and produce accidental hybrids. Length to 14 cm.

P. bimaculatus. Bali, Indonesia. Depth 45 m. Male, length 14 cm.

P. bimaculatus. Bali, Indonesia. Depth 45 m. Female, length 9 cm.

P. bimaculatus. Nuptial males. Lengths ~14 cm. Indonesia. **D** Bali. Gary Bell. **E** Pulau Putri, Java. Depth 10 m.

P. bimaculatus. Pulau Putri, Java, Indonesia. Depth 10 m. **F** male. Length 12 cm. **G** female, length 9 cm.

H

P. bimaculatus. Showing its nuptial banded pattern. South Java, Indonesia. D. 35 m. Male, length 12 cm.

I

P. bimaculatus. South coast Java, Java. Depth 35 m. Male, length 12 cm, with females, lengths about 9 cm.

Mirror Basslet *Pseudanthias pleurotaenia*

Anthias pleurotaenia Bleeker, 1857. Ambon.
Pseudanthias sheni Randall & Allen, 1989. WA.

Widespread Indo-West Pacific. Coastal to outer reefs at depths from 6 m down to moderate depths, but usually outer reefs along walls at about 25 m depth. Forms small schools, sometimes females and males in separate aggregations or mixed. Highly variable in colour, locally and geographically, and also in the size of the male's pale blotch (looks blue in natural light). This varies from being totally absent to almost half the body size, especially when showing nuptial colours. The presence or size seems to be in relation to the presence of other species living on the same reefs, or the density of populations, and in some populations only shows during spawning. Length to 16 cm.

A *P. pleurotaenia.* Flores, Indonesia. Depth 25 m. Nuptial male, ~16 cm.

P. pleurotaenia. Indonesia. **B** Flores. Male. Depth 35 m. Length ~15 cm. **C** Bali. Female. Depth 20 m. Length 13 cm.

D *P. pleurotaenia.* Qld, Australia. Male. Phil Woodhead.

E *P. pleurotaenia.* Flores, Indonesia. Depth 20 m. Female, ~14 cm.

F *P. pleurotaenia.* Sipadan. Depth 25 m. Male, ~16 cm. No blotch.

G *P. pleurotaenia.* Flores, Indonesia. Depth 25 m. Juvenile, 4 cm.

P. pleurotaenia. Indonesia. **H** Bali. Depth 25 m. **I** Flores. Depth 30 m. Males, length ~15 cm. Small and large blotch variations.

P. pleurotaenia. Flores, Indonesia. Depth 15 m. **J** female and **K** male, both length ~14 cm. Recovering population after destructive cyclone. No typical coloured males were present. It appeared that several females had change sex but did not follow up with colour.

P. pleurotaenia. **L** Qld, Australia. Male. D. 25 m. L. ~14 cm. **M** Bali, Indonesia. Male hybrid ×*bimaculatus*. Depth 20 m. L. 13 cm.

P. pleurotaenia. Milne Bay, PNG. Depth 45 m. Length 14 cm. One of a small group of similarly coloured individuals on a remote coral outcrop, whilst on the shallower reefs nearby males were darker and featured the more typical pale square on their sides.

P. pleurotaenia. Rowley Shoals, WA, Australia. Depth 50 m. Length 14 cm. Here, males can turn their pale square on and off.

P. pleurotaenia. Rowley Shoals, WA, Australia. Depth 40 m. Juveniles on drop-off with *Coniella* wrasses. Length about 55 mm.

Purple-yellow Basslet *Pseudanthias pictilis*

Anthias pictilis Randall & Allen, 1978. Lord Howe Island.

Eastern Australia from the Great Barrier Reef off Cairns to NSW, juveniles expatriating south to Montague Island; Lord Howe Island and New Caledonia. Adults on outer reef habitats to about 40 m depth. Juveniles in protected parts of rocky coastal and island bays on vertical walls with crevices or among boulders. Juveniles and females identified by their purple to yellow colouration. Males with distinctive caudal fin and unusual pale bar on the body. Length to 12 cm.

P. pictilis. Montague I., NSW, Australia. **C** & **D** Juveniles, lengths 30–40 mm. D. 25 m. **E** Female, 9 cm. D. 15 m.

P. pictilis. Male above, female below. Queensland, Australia. Depth 25 m. Lengths: **A** 12 cm. **B** 11 cm. Phil Woodhead.

P. pictilis. Male above, female below. Solitary Islands, NSW, Australia. Depth 25 m. Lengths: **F** 12 cm, and **G** 11 cm. Gary Bell.

P. randalli. Male. Flores, Indonesia. Depth 30 m. Length 65 mm.

Randall's Basslet *Pseudanthias randalli*

Anthias randalli Lubbock & Allen, 1978. Philippines.

Indo-Pacific, ranging from Indonesia to southern Japan. Clear coastal to outer reefs, usually on steep walls and at moderate depths to at least 70 m depth, but in some localities as shallow as 10 m. Juveniles and females pink with yellow snout and caudal fin. Males deep pink to red with longitudinal pattern of broad bands that may appear blue in deep water. Variable with habitat and location. In Bali, on black sand, the males are deep red with purple, whilst in Flores, on white sand, they are more pink with yellow. A small species, length to 7 cm, rarely 8 cm.

P. randalli. Males. Bali, Indonesia. **B** depth 25 m. Length 65 mm. **C** depth 15 m. Length 75 mm.

P. randalli. Male. Sipadan, Malaysia. D. 25 m. Roger Steene.

P. randalli. Female. Flores, Indonesia. D. 30 m. L. 6 cm.

P. randalli. Females. Flores, Indonesia. Depth 30 m. Lengths 6 cm.

Resplendent Basslet *Pseudanthias pulcherrimus*

Anthias pulcherrimus Heemstra & Randall, 1986. Maldives.

Maldives and Mauritius. Clear coastal to outer reefs, usually on steep walls and at moderate depths to at least 70 m depth, but in some localities as shallow as 10 m. Juveniles and females pink with yellow snout and caudal fin. Males deep pink to red with longitudinal pattern of broad bands that may appear blue in deep water. Variable with habitat and location. A small species, length to 7 cm, rarely 8 cm.

Remarks: very similar to *Pseudanthias randalli* and is possibly sub-specific or synonymous with it.

A

P. pulcherrimus. Male. Maldives

B

P. pulcherrimus. Male. Maldives. Jörg Aebi.

C

P. pulcherrimus. Female. Maldives

D

E

P. pulcherrimus. Maldives. Swimming against ceiling of cave **D** male. Scott Michael. **E** male and female.

Cave Basslet *Pseudanthias* cf *randalli*

Undetermined species.

Possibly a colour variation of *Pseudanthias randalli*, presently known only from a few localities in Palau, Saipan and Guam, Micronesia. Found deep along reef walls. Differs primarily in having bright yellow ventral fins, lacking a red posterior border. Length to 7 cm.

Remarks: similar form in Fiji, *Pseudanthias flavicauda* Randall & Pyle, 2001. Females of the species illustrated on this and the previous page are virtually identical in colour, whilst males somewhat more or less colourful between different regions. Besides colour there are no other clear differences between all these 'species' and they may represent geographical variants only.

A

P. cf *randalli*. Palau. Hiroshi Nagano.

B

P. cf *randalli*. Tinian. Depth 30 m. L. 7 cm. Yasuaki Miyamoto

A

White-bar Fairy-basslet
Pseudanthias leucozonus

Anthias (Pseudanthias) leucozonus.
Katayama & Masuda, 1982. Izu Peninsula, Japan.

Izu Peninsula region, including nearby islands, north to Hachijo Is off Tokyo. Adults occur on steep rock walls at depths of about 30–65 m, but juveniles often shallower. Usually seen solitary or in small groups mixed with other related planktivores. Readily identified by its white vertical bar centrally on the body. In general females are yellowish and males more reddish. In summer the area behind the white bar turns dark red on nuptial males. Length to ~13 cm.

B

P. leucozonus. **A** Izu, Japan. L. cm. Satoshi Yamamoto. **B** juvenile. Hachijo I., Japan. D. 15 m. L. 30 mm. Shigeru Harazaki.

C

D

P. leucozonus. Osezaki, Japan. Depth 55 m. **C** male left and female right. Lengths ~10 cm. **D** nuptial male. Length 12 cm. A. Mishiku.

Hawaiian Long-fin Basslet
Pseudanthias hawaiiensis

Anthias (Pseudanthias) ventralis hawaiiensis Randall, 1979. Hawaiian Islands.

Restricted to the Hawaiian and Johnston Islands. Reef habitats from about 30 m down. Juveniles sometimes shallower. Females yellow with pinkish flanks changing to orange with red or purplish spots when fully grown prior to sex change. Males with orange ventral fins and an orange band anteriorly on the anal fin, whilst the tail section turns pale. Length to 10 cm.

A

P. hawaiiensis. Hawaiian Is. Male with small female. Aquarium. Scott Michael.

B

C

P. hawaiiensis. Females. Hawaiian Is. Depth 25–30 m. John Hoover.

D

P. hawaiiensis. Hawaiian Island. Jim Watt (courtesy ANTHIS).

E

P. hawaiiensis. Male. Hawaiian Is. Depth 30 m. John Hoover.

Pacific Long-fin Basslet *Pseudanthias ventralis*

Anthias (Pseudanthias) ventralis ventralis
Randall, 1979. Pitcairn I.

Cook Islands to Pitcairn, South Pacific. Deep reef slopes. Male is mostly yellow over the back, except area below spinous dorsal which is orange to red, with deep-pink sides. Ventral fins long, but anal fin not greatly extended. Length to about 8 cm.

P. ventralis. Rarotonga, Cook Islands. **A** male. Kendall Clements. **B** male & female. Malcolm Francis.

Japanese Long-fin Basslet *Pseudanthias* cf *ventralis*-1

undetermined species

Ogasawara Is., Japan and Micronesia. Deep outer reef habitats, usually at depths over 40 m. Females yellow with pink flanks, developing purplish spots over the back with growth. Male with red spinous dorsal fin section and develops extremely long ventral and anal fin extensions. Length to 85 mm.

Remarks: shows similarities with *Pseudanthias hawaiiensis* and *P. ventralis*. Males lack the yellow or orange anal-fin band, prominent in other species of the complex.

P. cf *ventralis*-1. Male. Saipan. Depth 45 m. L. 6 cm. Yasuaki Miyamoto.

P. cf *ventralis*-1. **B** male. **C** young male. **D** female. **E** juvenile. Ogasawara Is, Japan. Yasuhiro Morita.

Coral Sea Long-fin Basslet *Pseudanthias* cf *ventralis*-2

undetermined species.

Great Barrier Reef and Coral sea. Found only on deep oceanic reefs in 40+ m depth. Occurs in small groups, usually with several females and a single male. In natural light in deep water these fish appear as blue and yellow, whilst the red markings look almost black. Males have a mostly red dorsal fin, ending with a red filament, and females have a red and blue line rising from top of eye, with red continuing above yellow sides and ending near caudal peduncle. A small species, to about 75 mm.

Remarks: a species of the '*ventralis*-complex' and is one of the most spectacular and beautiful of all the anthiid basslets.

A

P. cf *ventralis*. Male. L 6 cm. Great Barrier R., Australia. Phil Woodhead.

B

C

P. cf *ventralis*-2. Great Barrier Reef, Australia. **B** a young male just after changing sex. **C** male 6 cm. Phil Woodhead.

E

D

F

P. cf *ventralis*-2. Great Barrier Reef, Australia. Depth 45 m. **D** & **E** females. Length about 50 mm. **F** with male. Phil Woodhead.

Pacific Short-snout Basslet *Pseudanthias parvirostris*

Anthias (Mirolabrichthys) parvirostris Randall & Lubbock, 1981. Solomon Is.

Solomon Islands to Philippines and Indonesia west to Bali. Occurs in deep current-prone outer reef habitats from about 30 to at least 60 m depth. Usually seen in pairs or small groups. Female yellow-orange with purple stripe along dorsal-fin base, dividing on nape to above each eye and meeting again on snout. Males more orange and numerous small purplish spots on sides. Caudal fin reddish and outer margins broadly blue. Length to 8 cm.

P. parvirostris. Bali, Indonesia. D. 28 m. **A** male, length 6 cm. Takamasa Tonozuka. **B** female, 4 cm. Akira Ogawa.

Indian Short-snout Basslet *Pseudanthias* cf *parvirostris*

Undetermined species or subspecies.

Indian Ocean, known from Maldives and northern Sumatra to off northern Western Australia. Occurs on reef outcrops on moderately deep slopes, away from drop-offs, subject to strong currents. Whilst shallow in Sumatra, in other places usually at depths of 40 m or more. Females yellow, male more orange or red over the back. Caudal fin red with broad blue margin, the red looks black in deep water. A small species, reaching about 6 cm in length.

P. cf *parvirostris.* Pulau Weh, Sumatra. D. 20 m. L. 6 cm. Roger Steene.

P. cf *parvirostris.* Rowley Shoals, WA, Australia. Depth 57 m. Lengths 40 & 50 mm.

P. cf *parvirostris.* Maldives. Depth 42 m. Length of male 45 mm.

Yellow-backed Basslet *Pseudanthias bicolor*

Anthias (Mirolabrichthys) bicolor Randall, 1979. Hawaiian Islands.

Widespread Indo-Pacific, most tropical Pacific locations and as far west as Mauritius in the Indian Ocean. Deep coastal to outer reef slopes. Usually in small groups with remote bommies in current-prone areas. Male is unusual in having two yellow-tipped filamented rays heading the dorsal fin that are used for display. Juveniles are bright pink with ochre on upper part of the body from tip of the snout to upper caudal peduncle. Length usually to 11 cm, but to 13 cm in subtropical zones.

A

P. bicolor. Kerama, Japan. Depth 30 m. Male, length 12 cm.

B

P. bicolor. Bali, Indonesia. Depth 25 m. Juvenile, length 45 mm.

C

P. bicolor. Sangihe Is., Indonesia. Females.

D

P. bicolor. Bali, Indonesia. Depth 25 m. Female, length 8 cm. Gary Bell.

E

F

P. bicolor. Bali, Indonesia. Depth 25 m. Male, length 11 cm. **F** with parasite attached to side.

Lori's Basslet *Pseudanthias lori*

Anthias lori Lubbock & Randall, 1976. Loyalty Is.

Widespread West Pacific, ranging from southern Japan to northern Australia, and a few scattered populations in the South Pacific. Several localised colour forms. Common along outer reefs in Indonesia. Usually found along outer-reef walls with moderate currents at depths from 10–50 m. Young and females with series of orange-red blotches over the back below the soft part of the dorsal fin, followed by a red streak onto the tail. Length to 9 cm, excluding filamentous tips.

Remarks: report from Christmas Island, Indian Ocean, is based on *Pseudanthias flavoguttatus.*

A

P. lori. Bali, Indonesia. Depth 20 m. Male, length about 9 cm.

B

P. lori. Flores, Indonesia. Depth 30 m. Length 8 cm.

C

D

P. lori. **C** Flores, Indonesia. Depth 22 m. **D** Rowley Shoals, WA, Australia. Depth 30 m. Length 8 cm.

E

P. lori. Schooling females. Flores, Indonesia. Depth 25 m. Length 6 cm.

Red-saddled Basslet
Pseudanthias flavoguttatus

Anthias (Mirolabrichthys) flavoguttatus
Katayama & Masuda, 1980. Japan.

Restricted distribution from subtropical Japan, Philippines and along Wallace's Line to Bali and Christmas Island (Indian Ocean). Usually at depths over 30 m. Juveniles form large schools and are often mixed with other basslets when feeding on zooplankton. Looks similar to *Pseudanthias lori*, but saddles along the back are more pronounced and placed along entire dorsal fin to base of caudal fin. Lower lobe of caudal fin is often yellow. Length to 10 cm.

A

P. flavoguttatus. Bali, Indonesia. Depth 50 m. Length 75 mm. Akira Ogawa.

C

B

D

P. flavoguttatus. **B** Christmas Is., Indian Ocean. Depth 20 m. Length 45 mm. Roger Steene. **C** swimming with *P. lori* and *P. cichlops*. Bali, Indonesia. Depth 50 m. Length 25 mm. Akira Ogawa. **D** Oshima, Japan. Depth 36 m. Length about 30 mm. Yasuaki Miyamoto.

Central Pacific Basslet
Pseudanthias aurulentus

Anthias (Mirolabrichthys) aurulentus
Randall & McCosker, 1982. Line I., C. Pacific.

Central Pacific and appears to be restricted to the Line Islands and Cook Islands. Reef slopes and walls, where in small groups. A distinctive species with its red and white streaks along upper sides. Length to about 12 cm.

P. aurulentus. Aquarium. Specimens from Line Islands. Scott Michael.

Scribbled Basslet *Pseudanthias smithvanizi*

Anthias (Mirolabrichthys) smithvanizi
Randall & Lubbock, 1981. Solomon Is.

West Pacific, Indonesia to Marshall Islands, northern Australia and Cocos-keeling and Christmas Islands, Indian Ocean. In clear water along steep drop-offs with moderate currents. Usually at depths between about 10 and 50 m. Orange-pink, but appears more blue in deep water. It can be recognised by the yellow spots and scribbles along the body, and the large caudal fin. Sexes similar, males with long and filamentous caudal fin lobes. Often mixes with *Pseudanthias lori*, when schooling. Length to 90 mm.

A

P. smithvanizi. Bali, Indonesia. Depth 25 m. Length 9 cm.

B

P. smithvanizi. Christmas Island, Indian Ocean. Depth 25 m. Length 8 cm. Roger Steene.

C

P. smithvanizi. Flores, Indonesia. Depth 25 m. Pair, length 7–9 cm, male largest.

Flame Basslet *Pseudanthias ignitis*

Anthias (Mirolabrichthys) ignitis
Randall & Lubbock, 1981. Maldives.

Indian Ocean, Maldives and ranging east to southern Sumatra. Replaced further east by its sibling *Pseudanthias dispar* that occupies the same shallow habitat along reef edges. Males are recognised by the red caudal fin and red spot at base of pectoral fin. Females are plain orange with pinkish lines running from behind eye to pectoral-fin base. Length to 90 mm.

P. ignitis. Nuptial male. Maldives. Depth 10 m. Length 9 cm.

P. ignitis. Sumatra, Indonesia. Depth 10 m. Male, length 9 cm. Takamasa Tonozuka.

P. ignitis. Maldives. Depth 10 m. Length 9 cm. **C** displaying male. **D** male with cleaner wrasse and females.

P. ignitis. Colourful males. Maldives. Depth 10 m. Length 9 cm.

Fairy Basslet *Pseudanthias dispar*

Mirolabrichthys dispar Herre, 1955. Solomon Is.

West Pacific, Indonesia, Coral Sea and scattered Pacific locations as far as Line Islands. Usually found along clear water drop-offs with moderate currents at depths from about 6–10 m. Often occur in large schools of mixed sexes to feed on zooplankton when currents are sweeping the reef. Males display with a bright red dorsal fin and hold their ventral fins greatly extended. Females mainly orange with yellowish tail and a pink line from the tip of the snout angling down, passing through the lower part of the eye. Length to 90 mm.

P. dispar. Males. Great Barrier Reef. Depth 18 m. Gary Bell.

P. dispar. Bali, Indonesia. Depth 6 m. Schooling adults.

P. dispar. Bali, Indonesia. Depth 6 m. Length 9 cm. **D** unusually coloured caudal fin. **E** courting. **F** & **G** same individual, **G** yawning.

P. dispar. Bali, Indonesia. Depth 6 m. Length 9 cm. Male with unusually coloured caudal fin.

P. dispar. Flores, Indonesia. Depth 7 m. Length 8–9 cm. School with mixed sexes.

Sailfin Queen-basslet
Pseudanthias pascalus

Entonanthias pascalus
Jordan & Tanaka, 1927. Japan

Southern Japan to Sangihe Is. and northern Sulawesi, Indonesia. Inner reef channels and deep coral slopes in semi-protected habitats. Usually swims high above the substrate when feeding on zooplankton. Occurs mainly at depths of 10–25 m. Juveniles and females plain pink. Australian siblings have yellow spots. Males with red blotch on soft part of dorsal fin, often covering most of it, except for last few rays, and may extend as a band along spinous part. Length to at least 18 cm, excluding caudal fin filaments.

A

P. pascalus. Kerama, Japan. Depth 15 m. Length 18 cm.

B

C

P. pascalus. **B** Lembeh Strait, Sulawesi. Depth 6 m Length ~4 cm. **C** Iriomote, Japan. Depth 20 m. Length 16 cm.

D

P. pascalus. Kerama, Japan. Depth 15 m. Length ~18 cm.

Pacific Queen-basslet
Pseudanthias cf *pascalus*

South Pacific from Great Barrier Reef, Coral Sea to French Polynesia. Outer reef habitats. Occurs in small to large aggregations, depending on locality. Juveniles pink, developing a red line from tip of snout, running through eye, to pectoral fin base with growth and becoming yellow spotted. Males often with numerous dark spots on upper half of the body. Length to 15 cm.

Remarks: has been referred to as *Pseudanthias pascalus*, but it clearly is not that species.

A

P. cf *pascalus*. Male. Great Barrier Reef, Australia. Roger Steene.

B

C

P. cf *pascalus*. Males. **B** New Caledonia. Malcolm Francis. **C** Great Barrier Reef, Australia. Phil Woodhead.

D

E

P. cf *pascalus*. Females. Great Barrier Reef, Australia. Phil Woodhead.

F

P. cf *pascalus*. Schooling adults. Niue, South Pacific. Malcolm Francis.

P. tuka. Flores, Indonesia. Depth 6 m. Male, length 11 cm.

Purple Queen-basslet
Pseudanthias tuka

Mirolabrichthys tuka
Herre & Montalban, 1927. Philippines.

West Pacific, Indonesia, Philippines to Coral Sea and northeastern Australia. Schools swim above outer reef edges and along drop-offs to at least 30 m depth, but occur usually shallow. Large males are purple, but may look blue underwater, with a dark-red blotch in the dorsal fin that may look black in natural light. Females with yellow back that continues to the end of the caudal fin and lower part of that fin mirrored by a similar yellow streak. Length to 11 cm.

P. tuka. Bali, Indonesia. Depth 6 m. **B** female, length 8 cm. **C** changing sex. Length 9 cm.

P. tuka. Great Barrier Reef, off Cairns, Qld, Australia. Depth 15 m. Gary Bell.

P. tuka. Milne Bay, PNG. Depth 15 m. Length 10 cm.

P. tuka. Flores, Indonesia. Depth 20 m. Length 11 cm.

P. tuka. Bali, Indonesia. Depth 10 m. Length 11 cm.

P. tuka. Flores, Indonesia. Depth 25 m. Juveniles. Length 5 cm.

P. tuka. Flores, Indonesia. Depth 25 m. Schooling females. Lengths 6–9 cm.

A

P. cf *tuka.* Rowley Shoals, WA, Australia. Depth 9 m. Male, length 10 cm.

Yellow-fin Queen-basslet
Pseudanthias cf *tuka*

Rowley Shoals off north-western Australia. Forms large schools along the upper edge of outer reef walls at 10–20 m depth. Males deep-pink with a dark purple streak along base of soft part of dorsal fin with the rest of the fin orange to bright yellow. Juveniles and females with a yellow to ochre band over the back, continuing onto upper lobe of caudal fin and same coloured streak on lower lobe of caudal fin. Length to 10 cm.

Remarks: this species is clearly closely related to *Pseudanthias (Mirolabrichthys) tuka*, perhaps just a geographical variation, but the colour differences of males are quite dramatic.

B

C

P. cf *tuka.* Rowley Shoals, WA, Australia. Depth 9 m. **B** male, length 10 cm. **C** male and females, lengths 8–10 cm.

D

P. cf *tuka.* Rowley Shoals, WA, Australia. Depth 12 m. Female between two males, lengths 9–10 cm.

Yellow-back Queen-basslet
Pseudanthias evansi

Anthias evansi Smith, 1954. Kenya.

Widespread Indian Ocean, ranging east to Sumatra. Schooling species along upper parts of drop-offs and in outer reef lagoons. Feeds at various depths to near surface in pursuit of zooplankton. Readily identified by its pink (blue in natural light) and yellow colouration, similar to some of the fusiliers. Males tend to be more purplish in colour with numerous yellow spots in the dark areas. Length to 12 cm.

A

P. evansi. Maldives. Depth 10 m. Male, length 12 cm.

B

P. evansi. Maldives. Depth 15 m. Male, length 12 cm.

C

P. evansi. Maldives. Depth 6 m. Female, length 10 cm.

D

P. evansi. Maldives. Depth 9 m. Female, length 10 cm.

E

P. evansi. Pulau Weh, northern Sumatra. D. 10 m. Male, 9 cm.

F

P. evansi. Maldives. Depth 6 m. Length 11 cm.

Bartlett's Queen-basslet
Pseudanthias bartlettorum

Anthias (Mirolabrichthys) bartlettorum
Randall & Lubbock, 1981. Marshall Is.

Widespread Central Pacific, ranging to northern PNG to Line Islands and Micronesia. A schooling species on outer-reef slopes to about 30 m depth. Upper half of body and tail of juveniles and females bright yellow and lower part pink. Males more pink on sides to just below lateral line and have an elongated second dorsal fin spine. Length to 9 cm.

P. bartlettorum. **A** & **B** male. New Britain, PNG. Depth 20 m. Roger Steene. **C** Phoenix Is., Kiribati, Central Pacific. Jerry Allen.

P. bartlettorum. Phoenix Is., Kiribati, Central Pacific. Jerry Allen.

Marquesas Queen-basslet
Pseudanthias regalis

Anthias (Mirolabrichthys) regalis
Randall & Lubbock, 1981. Fatu Hiva, Marquesas.

Known only from the Marquesas. A schooling species on shallow rocky substrates to about 30 m depth. The various groups usually comprise similar sized individuals, including groups of males. Males are easily identified by their mostly pink body and a yellow patch on their head that extends onto the body. Females yellow along upper half of the body. Length to about 9 cm.

A

P. regalis. Male. Marquesas. John Hoover.

B

C

P. regalis. **B** groups of males and females with *Gymnothorax breedeni*. **C** females. Marquesas. John Hoover.

D

P. regalis. Schooling males. Marquesas. John Hoover.

Subfamily Anthiinae, Genus *Nemanthias* Smith, 1954

Masculine. Type species: *Nemanthias carberryi* Smith, 1954. A monotypic genus from the Western Indian Ocean. Has eleven dorsal spines, first two long and flexible, and lower lobe of caudal fin elongated in adults.

Threadfin Basslet *Nemanthias carberryi*

Nemanthias carberryi Smith, 1954. Kenya.

Widespread in the western Indian Ocean, east to the Maldives and south along the African coast with adult populations as far as the Aliwal Shoal. Juveniles may expatriate further south. A schooling species on upper reef slopes and near top of rocky reefs in vicinity of caves and crevices. Highly variable in colour and seems to mimic a number of other anthiid species as well as small fusiliers that also feed on plankton in open water, often swimming high above reefs. Juveniles and females plain yellowish pink. Males usually salmon-pink with numerous tiny yellow spots along scale rows, a yellow streak along base of soft part of dorsal fin and caudal fin with blue lobes. Variations may have bright yellow over the back, a bright yellow caudal fin, or a mostly bright red dorsal fin. Upper lip of adult protrudes into a fleshy tip, just like in the subgenus *Mirolabrichthys*. The elongated lower lobe of the caudal fin in large adults is occasionally damaged or bitten off, but re-grows. Length to 12 cm, excluding filamentous extension on caudal-fin lobes.

A

B

N. carberryi. Male. Maldives. Depth 8 m. Length 12 cm.

C

D

N. carberryi. Maldives. Variations, probably mimic colourations. Depth 7–10 m. Lengths about 10 cm.

N. carberryi. Maldives. Males with *Labroides* cleaner wrasses and females in background. Depth 10 m.

N. carberryi. Aliwal Shoal, South Africa. Depth 15 m. Length 12 cm.

Subfamily Anthiinae, **Genus** *Rabaulichthys* Allen, 1984

Masculine. Type species: *Rabaulichthys altipinnis* Allen, 1984. Comprises 3 small described species, 2 in the West Pacific and 1 in the western Indian Ocean, but more can be expected from different Pacific locations. The species are similar to *Luzonichthys*, but males have a very tall, sail-like dorsal fin, giving it the common name of 'sailfin basslet'. Females are rather plain. These fishes occur in small aggregations on moderately deep rubble slopes and, unless males display, are easily overlooked due to their small size and due to them blending into the background.

Rabaul Sailfin-basslet
Rabaulichthys altipinnis

Rabaulichthys altipinnis Allen, 1984.
Rabaul, New Britain.

Known only from off north-eastern Papua New Guinea, but is probably more widespread. Occurs on rubble slopes at depths of about 30 m or more. Feeds in small aggregations, usually swimming above the bottom in pursuit of zooplankton. Colour mostly plain pinkish, yellow or reddish above. Male's dorsal fin pale salmon-red without markings. Length 6 cm.

R. altipinnis. Male Holotype. Rabaul, New Britain. Jerry Allen.

Maldives Sailfin-basslet
Rabaulichthys stigmaticus

Rabaulichthys stigmaticus Randall & Pyle, 1989.
Maldives.

Maldive Islands and Sri Lanka. Occurs on rubble slopes with strong currents, where usually seen in small aggregations feeding just above the bottom on zooplankton. Females are plain, but males are readily identified when displaying with their tall dorsal fins, the soft part of which is pale-blue with an elongate black spot in its centre. Length 6 cm.

R. stigmaticus. Male. Maldives. Depth 45 m. Length 6 cm.

Philippines Sailfin-basslet
Rabaulichthys cf *suzukii*

Undetermined species. Philippines.

Specimens were collected in the Philippines and it seems similar to *Rabaulichthys suzukii*. Possibly the same species or a geographical colour variation, but it could also represent a new species. Differs from the Japanese form in males having bright red ventral fins. Length about 6 cm.

A

B

R. cf *suzukii.* Male in aquarium. From Philippines. Hiroyuki Tanaka.

Suzuki's Sailfin-basslet
Rabaulichthys suzukii

Rabaulichthys suzukii Masuda & Randall, 2001. Japan.

Izu Peninsula, Japan, probably ranging to the Ryukyu or Ogasawara Islands. Occurs moderately shallow on the Izu Peninsula at 5–15 m depth and probably occurs deeper in more tropical zones. Very similar to *Rabaulichthys altipinnis*, but male with lavender dorsal and anal fins, and deeper-forked caudal fin. Length to 7 cm.

A

B

C

R. suzukii. Kochi, Japan. **B** nuptial male. **C** female. Tomonori Hirata.

D

E

R. suzukii. Displaying males. Kochi, Japan. Tomonori Hirata.

Subfamily Anthiinae, Genus *Tosana* Smith & Pope, 1906

Feminine. Type species: *Tosana niwae* Smith & Pope, 1906. A monotypic genus from the Western Pacific. Features a slender body, lateral-line high on sides with 2 scale rows between it and middle of spinous dorsal fin. Median rays of pectoral fin unbranched. Outer rays of caudal fin and second ray of ventral fin greatly prolonged, especially in adults.

Slender Fairy-basslet
Tosana niwae

Tosana niwae Smith & Pope, 1906. Tosa, s. Japan.

Southern Japanese water and China seas, but also found off the east coast of Australia from southern Queensland to central NSW. Deep rocky reefs, occasionally trawled and known range 86–130 m. Mostly pink, with a yellow mid-lateral band at eye level. Length to about 16 cm excluding caudal-fin filamentous parts.

T. niwae. NSW, Australia. Depth ~70 m. Length 12 cm. Ken Graham.

SUBFAMILY ANTHIINAE, GENUS *Luzonichthys* Luzonichthys, 1954

Masculine. Type species: *Mirolabrichthys waitei* Fowler, 1931. A small group of fairy basslets in which 6 species are recognised in the tropical Indo-Pacific. Found on coral reefs, usually forming large schools, feeding on zooplankton in the water column. They are small, reaching less than 70 mm in length, and show no obvious differences between sexes. Unlike other basslets, that usually have a continuous dorsal fin, the spinous and soft parts are separate and sometimes these fishes are referred to as 'splitfin basslets'.

Right, below and next page:
Luzonichthys sp.
Great Barrier Reef, off Cairns. This page Phil Woodhead, next page Roger Steene.

Like many other populations, this species is presently included with *Luzonichthys waitei*, and it seems that this taxon is used as a 'catch-all' for those with similar morphology and meristic values, whilst ignoring different colours. The Queensland population looks quite different from those in the Philippines (type-locality) and Indonesia. Often forms large schools on the outer-reef edge of the Great Barrier Reef when the water is rich with plankton during the out-going tides, usually mixed with various other small planktivores, as shown below and on next page.

L. microlepis. Maldives. Length 55 mm. Jörg Aebi.

Smith's Pygmy-basslet
Luzonichthys microlepis

Naurua microlepis Smith, 1955. Aldabra.

Western Indian Ocean Islands from Aldabra to the Maldives. Outer reef habitats but few specimens are known and it is rarely observed. Orange-red above and pink to blue below. Red streaks along caudal fin lobes. Length to 7 cm.

Remarks: virtually identical in colour as well as morphology to *Luzonichthys earlei* that occurs in the Hawaiian Islands.

A

L. taeniatus. Flores, Indonesia. Depth 30 m. Length 35 mm.

Orange-striped Pygmy-basslet
Luzonichthys taeniatus

Luzonichthys taeniatus
Randall & McCosker, 1993. Indonesia.

Indonesia, Banda and Flores Seas. Occurs in small schools along the bases of deep drop-offs at depths of 30 m or more. Bright pink with orange from the snout, over head and back to below soft dorsal fin. A small species, length to about 40 mm.

Remarks: previously reported as *Luzonichthys earlei* which is very similar in colour.

B

L. taeniatus. Flores, Indonesia. Depth 30 m. Length 35 mm.

Purple Pygmy-basslet *Luzonichthys waitei*

Mirolabrichthys waitei Fowler, 1931. Philippines.

Widespread Indo-West Pacific, but some populations may represent different species. True species occurs in the Western Pacific, from Philippines and Indonesia to Papua New Guinea and northern Coral Sea. Common in Indonesia along outer reef slopes on the upper edge of steep walls to depths of about 20 m, usually forming large aggregations during feeding. Pink or purplish with dusky to ochre back, grading to yellow or orange on caudal peduncle. Caudal-fin lobes pink to yellow or with yellow tips. Length to about 50 mm.

A

L. waitei. Izu, Japan. D. 7 m. L. ~45 mm. Yasuaki Miyamoto.

B

C

D

L. waitei. Flores, Indonesia. Depth 10 m. Length 50 mm. Note: profile of dorsal fin in **D**, typical for genus.

Whitley's Pygmy-basslet *Luzonichthys whitleyi*

Naurua whitleyi Smith, 1955.
Replacement name for *N. waitei* Whitley & Colefax, 1938.
Gilbert Islands.

Known from various localities in the Central Pacific and seems to be widespread. Occurs at moderate depths in small groups, usually in excess of 20 m. Identified by the bright yellow snout that extends as a stripe over the back to below soft dorsal fin. Length to 50 mm.

Remarks: very similar or identical to *Luzonichthys williamsi* that occurs in the Loyalty Islands.

L. whitleyi. Palau. Ryoichi Satoh.

Subfamily Anthiinae, Genus *Tosanoides* Kamohara, 1953

Masculine. Type species: *Tosanoides filamentosus* Kamohara, 1853. Comprises 2 small species, primarily known from subtropical Japanese waters, inhabiting deep reefs, usually at depths of more than 50 m. Cave dweller, usually seen solitary, but due to deep-water habitats it is rarely seen by divers and only in a few locations. Both species are presented here with all photographs taken in the same cave during a single dive, as well as *Odontanthias borbonius* (next genus, page 92), taking 21 photographs in less than 10 minutes at 60 m depth and all in focus, quite a memorable dive!

A

Thread-tail Cave-basslet
Tosanoides filamentosus

Tosanoides filamentosus Kamohara, 1853. Japan.

Subtropical Japan, mainly Pacific side of mainland. Deep rocky reefs. Lives in caves, usually in depths over 50 m. Rarely observed due to its deep water preference and small size. Pink with yellow markings, its first dorsal-fin spine distinctly longest which readily distinguishes it from the otherwise very similarly coloured female *Tosanoides flavofasciatus*. Length to 9 cm.

B

C

T. filamentosus. Male. Oshima, Japan. Depth 60 m. Length 65 mm.

Candy-striped Cave-basslet
Tosanoides flavofasciatus

Tosanoides flavofasciatus
Katayama & Masuda, 1980. Oshima, Japan.

Primarily known from the Izu region of Japan where it can be observed by diving at depths of 50 m or more. Also reported from the Tonga Ridge in the Pacific. This is a stunningly beautiful species, males with bright pink and yellow longitudinal stripes. Females pink with a small yellow spot on each scale forming lines along vertical scale rows, and similar to *Tosanoides filamentosus,* but is readily distinguished from it by the short first dorsal-fin spine. Length to 9 cm.

A

T. flavofasciatus. Male. Oshima, Japan. Depth 60 m. Length 8 cm.

B

T. flavofasciatus. Oshima, Japan. D. 55 m. L. ~3 cm. Yasuaki Miyamoto.

C

D

T. flavofasciatus. Male. Oshima, Japan. Depth 60 m. Length 8 cm.

Subfamily Anthiinae, **Genus** *Odontanthias* Bleeker, 1873

Masculine. Type species: *Serranus borbonius* Cuvier, 1828. Indo-Pacific. Comprises at least 11 deep-dwelling species that mostly occur in depths of 100 m+ and are rarely seen, usually caught on fishing lines from rocky reefs. Colourful deep-bodied species with pink red or yellow, and with large fins that are often filamentous. Males and females are different in colour and fin morphology, and in some species the sexes may go under different names. Most species range from 10–20 cm in length, but some attain about 30 cm. Some species are successfully kept in public aquariums in Japan and Hawaii, specimens that were caught in traps or on fishing lines. *Odontanthias* is often wrongly synonymised with the Atlantic genus *Holanthias*.

1 *Odontanthias borbonius* (Valenciennes, 1828) Réunion I. 92
2 *Odontanthias caudicinctus* (Heemstra & Randall, 1986) Kenya 95
3 *Odontanthias chrysostictus* (Günther, 1872) Indonesia 94
4 *Odontanthias dorsomaculatus* Katayama & Yamamoto, 1986 W. I.O. n.i.
5 *Odontanthias elizabethae* Fowler, 1923 Hawaiian Is. n.i.
6 *Odontanthias flagris* Yoshino & Araga, 1975 Japan n.i.
7 *Odontanthias fuscipinnis* (Jenkins, 1901) Hawaiian Is. 95
8 *Odontanthias kingyo* (Kon, Yoshino & Sakurai, 2000) Japan n.i.
9 *Odontanthias natalensis* (Fowler, 1925) Natal n.i.
10 *Odontanthias rhodopeplus* (Günther, 1872) Indonesia n.i.
11 *Odontanthias unimaculatus* (Tanaka, 1917) Japan 94
n.i. = not included

Blotched Seaperch *Odontanthias borbonius*

Serranus borbonius Valenciennes, 1828. Réunion I.

Subtropical southern Japan, Ogasawara Is., and ranging south to Taiwan. Seemingly separate population in the Western Indian Ocean, Réunion and off Durban, South Africa. Occurs on deep reefs, usually at depths of 60 m or more. Sometimes caught on fishing lines. A spectacular basslet, mostly pink with greenish yellow blotches on the body and stripes behind and below eye over cheek. Filaments on soft dorsal and caudal fin rays, and third dorsal-fin spine elongated in large adults. Length to at least 15 cm.

O. borbonius. Oshima, Japan. Depth 60 m. Length 12 cm.

O. borbonius. Oshima, Japan. Depth 60 m. Length 12 cm.

D

O. borbonius. Guam. Large specimen from a depth of 150 m. Length 15 cm. Robert Myers.

E

O. borbonius. Oshima, Japan. Depth 60 m. Length 12 cm.

O. katayamai. Male. Guam. Depth ~180 m. L. 18 cm. Robert Myers.

Katayama's Seaperch
Odontanthias katayamai

Holanthias katayamai
Randall, Maugé & Plessis, 1979. Guam.

Mariana Islands. Deep rocky reefs to about 300 m depth. Body mostly pink with yellow over head to below end of spinous dorsal fin. Series of yellow spots on sides forming diagonal lines, Most soft rays of dorsal fin with long filaments, the one on second soft ray distinctly longest. Ventral and anal fins with long white filaments. Length to 18 cm.

Remarks: paratypes used in the original description included specimens of the next species. However, that differs in colour and morphology. Most obviously: it lacks multiple filaments on the dorsal fin, has a conspicuous yellow spot low on cheek, and larger eyes.

A

O. chrysostictus. Okinawa, Japan. Depth ~100 m. L. 16 cm. Masahiro Tarui.

Yellow-spot Seaperch
Odontanthias chrysostictus

Holanthias chrysostictus Günther, 1872.
Manado, Sulawesi.

Tropical southern Japan ranging to Indonesia. Deep rocky reefs, to about 200 m depth. Body pink to purple, yellow dorsally, with yellow scribbled lines and spots, and band from snout to pectoral fin. In adults he anterior second soft ray on dorsal, anal and ventral fins are extend into long filaments. Length to 20 cm.

O. chrysostictus. Subadult. Oshima, Japan. Depth about 65 m. Length ~10 cm. **B–D** Osamu Hoshino.

Black-tip Seaperch *Odontanthias unimaculatus*

Anthias unimaculatus Tanaka, 1917. Wakayama Pref., Japan.

Pacific coast of southern Japan, and ranging south to Taiwan. A rare species that is occasionally caught on lines from deep rocky reefs. Body pinkish with pearly spots and head with yellow around and band below eye from snout to operculum edge. Third dorsal fin spine long with black membrane at tip (shown as spot in dorsal fin when spines laid back, as shown in photograph. Length to 20 cm.

O. unimaculatus. Tokai aquarium, Japan. Length 15 cm.

Yellow-finned Seaperch *Odontanthias fuscipinnis*

Anthias fuscipinnis Jenkins, 1901. Hawaiian Islands.

Hawaiian Island region. Usually occurs very deep, shallowest depth of capture known is about 50 m, near rocky reefs. Juveniles mainly yellow, adults pink with bright yellow fins and black markings near their base. Median fins with blue margins and several filaments on the anterior soft rays of the dorsal fin. Length to about 24 cm.

O. fuscipinnis. Aquarium. Length 16 cm. Robert Myers.

Black-blotch Swallowtail *Odontanthias caudicinctus*

Holanthias caudicinctus Heemstra & Randall, 1986. Kenya.

A deep water species known only from off Kenya at 140–300 m depth. Females plain, pinkish, orange above and pale below, and a yellow stripe from tip of snout over cheek and pink stripe above running over top of eye. Males with a dark blotch and a white saddle-like marking on caudal fin-base and peduncle. Small species, the most slender in genus, length to about 14 cm to fin tips.

A

O. caudicinctus. Female. L. 60 mm SL. Phil Heemstra.

B

O. caudicinctus. Male. Holotype. Length 92 mm SL. Phil Heemstra.

SUBFAMILY ANTHIINAE, **GENUS** *Sacura* Jordan & Richardson, 1910

Feminine. Type species: *Anthias margaritaceus* Hilgendorf, 1879. A small genus with at least 4 species, usually living in deep water or offshore. All have rather deep bodies (40–50% of standard length) and adults feature long filamentous rays on dorsal and caudal fins. Only one species is well known and commonly observed in Japanese waters and is a beautiful fish that is often on display in public aquariums in Japan.

Cherry Seaperch *Sacura margaritacea*

Anthias margaritaceus Hilgendorf, 1879. Japan.

Subtropical southern Japan and Korean waters. Forms schools on deep rocky reefs subject to strong currents, at depths of about 40–50 m, often rising high above the reef to feed on zooplankton. Occurs shallower but adults rarely at depths less than 20 m. Juveniles usually form small aggregations in large caves. Juveniles are recognised by their plain yellow to pink deep body and a large black spot centrally on the dorsal fin on a section with 3–4 spines. Males scarlet red with white blotches and long filaments on the dorsal and caudal fins. Length to 15 cm.

A *S. margaritacea.* Male. Izu Oceanic Park, Japan. Depth 35 m. Length 15 cm.

B *S. margaritacea.* Male. Oshima, Japan. Depth 25 m.

C *S. margaritacea.* Male. Osezaki, Japan. Depth 30 m.

D

E

S. margaritacea. **D** female. Osezaki. **E** male. Still showing dorsal spot. Izu Oceanic Park, Japan.

F

S. margaritacea. Izu Oceanic Park, Japan. Above: depth 25 m. Male (centre) and female (*Scorpaenodes littoralis* in background). Below: depth 30 m. Juveniles in cave with *Labroides dimidiatus* cleaner wrasse and *Pseudanthias fasciatus* in background.

G

SUBFAMILY ANTHIINAE, GENUS *Serranocirrhites* Watanabe, 1949

Masculine. Type species: *Serranocirrhites latus* Watanabe, 1949. A monotypic genus in the Western Pacific. A small deep-bodied species that mainly dwells in caves on outer coral reefs. It features short spines in the dorsal fin, but has a large soft-rayed section with 18–20 rays, a large forked caudal fin with extended tips, and anal fin is pointed, ending with filamentous tip. Pectoral fin long, reaching to well above anal fin and some rays extended with filaments.

A

S. latus. Sulawesi, Indonesia. Depth 35 m. Length 8 cm.

Swallow-tail Basslet
Serranocirrhites latus

Serranocirrhites latus Watanabe, 1949. Japan.

Widespread West Pacific, known from southern Japan, Indonesia, northern Australia and Coral Sea. Primarily occurs in clear water along drop-offs at moderate depths, of about 20–70 m, and usually in caves, sometimes swimming upside-down. Mainly a solitary species, but where common it occasionally forms loose aggregations. A deep bodied species with large forked tail, pointed anal fin and elongated rays on pectoral fins. Colour rose-pink with yellow markings, stripes on head, spots on body and yellow streaks in caudal fin. Length to about 85 mm, excluding long tail-lobes.

B

S. latus. Oshima, Japan. D. 36 m. L. ~25 mm. Yasuaki Miyamoto.

C

D

S. latus. Flores, Indonesia. Depth 35 m. Lengths 5 & 6 cm.

SUBFAMILY ANTHIINAE, GENUS *Caprodon* Temminck & Schlegel, 1843

Masculine. Type species: *Anthias schlegeli* Günther, 1859. Proposed originally without species, and species added by Günther who treated it as a synonym of *Anthias*. A temperate to subtropical genus in need of revision, comprising a few species from scattered localities of the Pacific rim. A moderately large anthiid that lives on deep reefs. Normally caught on fishing lines or by trawlers, but may rise to shallow depths in pursuit of zooplankton and may be observed by scuba divers at depths of about 50 m or less. It has a low spinous dorsal and rounded, truncate or concave caudal fin. Pectoral fin long and pointed with upper rays short and progressively increasing in length to central rays and lower half more rounded. Lower jaw usually clearly protruding from upper. These fishes are rarely kept in captivity.

Schlegel's Seaperch *Caprodon schlegeli*

Anthias schlegeli Günther, 1859. Japan.

Known from Japan and Taiwan. Primarily occurs on rocky reefs on continental shelves. In Japan it enters relatively shallow depths and is regularly observed by divers, usually swimming in open water above reefs with other planktivores. Usually caught on lines or trawled from depths in excess of 100 m. Females pinkish with dark blotches on back and base of dorsal fin. Males with yellow markings and large black area on spinous dorsal fin. Length to 45 cm.

Remarks: reports from Australia, Hawaii and Chile appear to be based on other species.

A

C. schlegeli. Tokai Aquarium, Japan. Length ~30 cm.

B

C

C. schlegeli. Juveniles. Izu Peninsula, Japan. Length ~8–10 cm. Satoshi Yamamoto.

D

C. schlegeli. Male. Izu, Japan. Depth about 55 m. Length ~35 cm. Shouichi Kato.

Sunrise Seaperch
Caprodon krasyukovae

Caprodon krasyukovae
Kharin & Dudarev, 1983.
Lord Howe Rise, off NSW, Australia.

Lord Howe Island region to southern Qld and NSW coasts. Occurs on deep offshore reefs at depths over 50 m. Juveniles and females yellow-pink with dark bands, eyes yellow, and caudal fin lunate. Bands fade with growth and males marked with yellow: behind eye, over back, and on spinous dorsal and caudal fin. In the latter the median 6 or 7 rays elongate with growth. Length to 46 cm.

A

C. krasyukovae. Male. Lord Howe Is. D. 300 m. L. 43 cm. Carl Bento, Australian Museum.

C. krasyukovae. NSW. Australia. D. 140–150 m. Lengths: **B** 18 cm, **C** 43 cm, **D** 28 cm, **E** 35 cm, **F** 38 cm. Ken Graham.

Hawaiian Seaperch
Caprodon unicolor

Caprodon unicolor Katayama, 1975.
Midway I., n. Pacific. 170 m.

Only known from Midway and the Hawaiian Islands where on deep reefs. Forms small to moderately large schools. Readily identified by the the pink colour, that appears blue in deep water, and the barred pattern on the posterior two-third of the body and shape of the caudal fin. Length to at least 30 cm

C. unicolor. Hawaiian Islands. Depth 35 m. Length 25 cm. John Hoover.

Longfin Seaperch *Caprodon longimanus*

Anthias longimanus Günther, 1859. New Zealand.

A common species around the offshore islands of northern New Zealand and also occurs in NSW, Australia. In New Zealand females form large schools in open water and may rise to the surface in pursuit of krill, but usually found at depths of between 15–60 m. Males are much rarer and usually seen singly. In Australia mainly in deep water and usually caught on lines from about 100 m depth, occasionally seen shallower on offshore reefs or islands. Small juveniles secretive in shallow protected reefs. These are yellow and pink with dark head. Females plain pink and males adorned with yellow spots and dark blotch on soft dorsal fin. Length to 50 cm, but usually much smaller.

Remarks: this species has a much more concave caudal fin than its congeners. Juveniles and females are much like *Caesioperca*.

C. longimanus. NSW, Australia. Male. Montague Island. Depth 30 m. Length 35 cm.

C. longimanus. **D** Narooma, NSW. Depth 3 m. Length 35 mm. **E** female. New Zealand. D. 20 m L. 24 cm. Malcolm Francis.

C. longimanus. NSW, Australia. Caught on line. **F** female from 50 m. Length 25 cm. **G** male from 80 m. Length 36 cm.

Subfamily Anthiinae, **Genus** *Lepidoperca* Regan, 1914

Feminine. Type species: *Lepidoperca inornata* Regan, 1914. Comprises 10 species, one of which in the southern Atlantic, and the rest from Australian and new Zealand waters. Most species are only known from deep shelf waters to several hundreds of metres depth. Rarely in less than 60 m, and most species are only known from trawling. They are planktivores and likely swim in close vicinity to reefs. Some species are known from few specimens only and these may remain closer to the bottom to feed or swim off vertical reef walls that are avoided by trawlers.

1 *Lepidoperca aurantia* Roberts, 1989	New Zealand	6 *Lepidoperca inornata* Regan, 1914	New Zealand
2 *Lepidoperca brochata* Katayama & Fujii, 1982	Se Australia	7 *Lepidoperca magna* Katayama & Fujii, 1982	Norfolk Ridge
3 *Lepidoperca caesiopercula* (Whitley, 1951)	Qld, Australia	8 *Lepidoperca occidentalis* Whitley, 1951	Western Australia
4 *Lepidoperca coatsii* (Regan, 1913)	S Atlantic	9 *Lepidoperca pulchella* (Waite, 1899)	E Australia
5 *Lepidoperca filamenta* Roberts, 1987	S Western Australia	10 *Lepidoperca tasmanica* Norman, 1937	Tasmania

L. pulchella. Adult specimen. Sydney markets. Length 22 cm.

Eastern Orange Seaperch
Lepidoperca pulchella

Anthias pulchellus Waite, 1899. Wollongong, NSW.

East coast of Australia from northern NSW to Tasmania, also New Zealand, and a very similar species, *Lepidoperca filamenta* in south-western Australia. A schooling species near reefs, usually deep shelf habitats at depths of about 100–300 m. Females plain orange and males with black spot anteriorly on soft dorsal fin. Western species has an additional small black spot centrally on pectoral and anal fins. *L. pulchella* reaches a length of 28 cm, whilst *L. filamenta* reaches 18 cm.

Lepidoperca brochata. NSW, Australia. D. ~160 m. L. 25 cm. Ken Graham.

Fang-tooth Seaperch
Lepidoperca brochata

Lepidoperca brochata Katayama & Fujii, 1982. South-eastern Australia.

Known only from off NSW, south-eastern Australia. Continental shelf, usually at depths of 200–300 m, but occasionally shallower. Bright pink with yellow longitudinal streaks on sides and yellow fins, and iris white. Length to 28 cm.

Lepidoperca occidentalis. Trawled off Western Australia. Length 18 cm.

Slender Orange Seaperch
Lepidoperca occidentalis

Lepidoperca occidentalis Whitley, 1951. Off Perth, Western Australia.

South-western Australia, ranging from Port Lincoln, S.A. to the Houtman Abrolhos, W.A. Deep shelf waters at depths of 60–200 m. Probably schools close to rocky reefs as this species is only known from a few trawled specimens. Identified by its slender body. Females plain reddish pink above and pale below and thin black posterior margin on operculum. Males with black spot on soft dorsal fin and broad yellow margin on spines and anterior soft part of anal fin. Length to about 20 cm.

Wavy-line Seaperch
Lepidoperca tasmanica

Lepidoperca tasmanica Norman, 1937.
Tasmania.

Southern Tasmanian and New Zealand waters. Normally found at depths over 150 m, but enters much shallower water in Fiordland, southern New Zealand, where it occurs along steep walls, as shallow as 20 m. Distinctive in having red wavy lines on sides along horizontal scale-rows. Length to at least 20 cm.

Remarks: Originally described from Tasmania and for a long time was thought to be a juvenile stage of *Lepidoperca pulchella*, but recently it was re-instated as a valid species.

A

B

L. tasmanica. Fiordland, southern New Zealand. Length ~ 15 cm. Quentin Bennett.

Sharp-head Seaperch
Lepidoperca magna

Lepidoperca magna Katayama & Fujii, 1982.
Norfolk ridge, South Pacific.

Temperate south-west Pacific from off NSW, Australia to the Lord Howe and Norfolk Island regions, and south to New Zealand. Usually trawled in depths over 150 m. A deep-bodied species and head profile is concave over eye. Deep pink with a distinct yellow stripe over the lateral-line. Length to 25 cm.

L. magna. NSW, Australia. Depth 460 m. Length 19 cm. Ken Graham.

Short-head Orange Seaperch
Lepidoperca caesiopercula

Anthias pulchellus caesiopercula Whitley, 1951.
Off near Cape Moreton, Qld, Australia.

Eastern Australia from southern Queensland to central New South Wales. Occurs on inshore reefs, but moderately deep at 50+ m depth. A robust species with relatively short head and low spinous dorsal fin. Plain orange-red with yellow soft dorsal and caudal fins. Much smaller than most congeners, length to 15 cm.

L. caesiopercula. NSW, Australia. Depth 55 m. Length 12 cm. Ken Graham.

Subfamily Anthiinae, **Genus** *Caesioperca* Castelnau, 1872

Masculine. Type species: *Serranus rasor* Richardson, 1839. A temperate genus from southern Australian and New Zealand waters, comprising 3 species. Similar to *Caprodon* & *Lepidoperca*, but can be distinguished from them by its shorter and more evenly rounded pectoral fin. They feed primarily on zooplankton or small swimming crustaceans such as mysids and krill, but accept most foods in captivity and are relatively easy to keep if kept cool and regularly fed. Due to size, only suitable for a large home aquarium or public aquarium.

Barber Perch *Caesioperca rasor*

Serranus rasor Richardson, 1839.
Port Arthur, Tasmania.

Australia's south coast, west to Albany, northern Tasmania and in Victoria east to Wilsons Promontory. Often in large schools feeding in strong currents above remote rocky reefs or shipwrecks. Juveniles enter estuaries. Small juveniles are pink with a mostly black head, changing to plain pink all over with growth. Males covered with blue and yellow markings and usually have an obvious black mark on their side that varies in allocation and from a blotch or vertical bar. Length to 25 cm.

A

C. rasor. Port Phillip Bay, Victoria, Australia. Depth 15 m. Length 22 cm.

B

C

C. rasor. Port Phillip Bay, Victoria. **B** depth 15 m. Male. Length 24 cm. **C** depth 3 m. Length 35 mm.

D

E

C. rasor. Port Phillip Bay, Victoria. Depth 15 m. **D** large female. Length 20 cm. **E** juveniles. Lengths about 6–7 cm.

The individual on the right had a ghostly white appearance when seen underwater. It was swimming with several normally coloured individuals and a morph seen only once. It stood out from the others and may have been targeted by predators often, as it was extremely flighty for this species (which is normally easily approached). It was not possible to get a close photograph of this individual, but even the extensivily cropped image shows the strange 'koi carp-like' colour morph.

F

C. rasor. Kent Group, Bass Strait, Tasmania. Depth 25 m. Length 20 cm.

G

C. rasor. Kent Group, Bass Strait, Tasmania. D. 25 m. Schooling adults, females (mostly pink) and males (with many blue markings).

Red-lined Seaperch
Caesioperca sp 1

Caesioperca n sp
Rottnest I., Western Australia.

Only known from subtemperate waters of Australia's west coast from Cape Leeuwin to Rottnest Island. Occurs in schools on off-shore reefs at moderate depths of 20 m or more. Identified by the stripes on its sides that look grey or black in natural light. Length to 15 cm.

C. sp 1. Rottnest Island. Western Australia. Length 15 cm. Barry Hutchins.

C. lepidoptera. Port Phillip Bay, Victoria, Australia. Depth 30 m. Length 25 cm.

Butterfly Perch *Caesioperca lepidoptera*

Epinephelus lepidopterus
Bloch & Schneider, 1801. New Zealand seas

Temperate waters of Australia and New Zealand. A common species in southern Australian waters where often clouding the water above offshore reefs. Usually most abundant on reefs subject to strong oceanic or tidal currents. Juveniles in small groups, staying within the shelter of reefs. Plain pink when young, pale on belly, developing a dark blotch mid-laterally above the anal fin. Adults with a bluish streak behind upper half of eye and some individuals become covered with small dusky spots. Length in Australia to about 30 cm, reported in New Zealand to 40 cm.

B
C. lepidoptera. Hobart, Tasmania. Depth 20 m. Length 30 cm.

C
C. lepidoptera. Sydney, Australia. Length 45 mm.

D
C. lepidoptera. Bicheno, Tasmania. Depth 25 m. Juveniles, length about 6 cm.

C. lepidoptera. NSW, Australia. **E** Montague I. Depth 25 m. Subadult, length 12 cm. **F** Bermagui. Depth 30 m. Strong current habitat.

On right; a strangely coloured individual lacking the typical black blotch, thought to represent a different species. It was found swimming amongst hundreds of others with the blotch (as shown in background, and the picture below). Also, normally there is no black pigmentation in front of and on the tail, and has clear pectoral fins. Not shy, but with strong currents running, it was hard work to get close for a photograph and, after finally succeeding in getting a good picture, it turned around and .... it was normally coloured on the other side!

Similar variations have been observed and photographed in other anthiid planktivorous species, but usually the colouration is the same on both sides. However, species often mix and may even change colour to blend in with other species. This is especially the case with those that are rare in a certain area feeding amongst local planktivores.

C. lepidoptera. Variation. Bermagui, NSW, Australia. Depth 25 m. Length 20 cm.

C. lepidoptera. Bermagui, NSW, Australia. Depth 30 m. Schooling adults about 25 cm long.

Subfamily Anthiinae, **Genus** *Plectranthias* Bleeker, 1873

Masculine. Type species: *Plectropoma anthioides* Günther, 1871. Comprises about 45 species of mostly small, little known species, many of which are known from just 1 or 2 specimens. Most species live in deep water, are usually very small (less than 10 cm) and in addition they are very secretive. These perchlets are never or rarely seen by scuba-divers and therefore good photographs of only a few species are available and included here. The few species that occur in shallow water have blotched patterns that blend in with the background and live amongst corals or in narrow crevices. They are often mistaken for juvenile hawkfishes (Cirrhitidae) because of their general appearance and the way they rest on the bottom. However, hawkfishes have small tufts at the tips of the dorsal-fin spines, even at small sizes.

Little is known about these fishes in captivity, but as they appear to feed on small crawling crustaceans they would be well suited for the reef-aquarium. Because of their generally pygmy size, these perchlets can be kept in small home aquariums, but they have a large mouth and companion fishes should not be much smaller, in case they are seen as potential prey.

Plectranthias longimanus. An adult pair, photographed at 15 m depth in Milne Bay, Papua New Guinea. This is one of the few species in the genus commonly observed. These little anthiid fishes, less than 35 mm long, look just like juvenile hawkfishes in the way they perch themselves on corals, but locations are usually more hidden. Other species are much more secretive and the majority of species have not been observed in the wild, but were collected with the use of chemicals or trawled from deep water.

1	*Plectranthias alleni* Randall, 1980	n.i.
2	*Plectranthias altipinnatus* Katayama & Masuda, 1980	109
3	*Plectranthias anthioides* (Günther, 1872)	n.i.
4	*Plectranthias azumanus* (Jordan & Richardson, 1910)	113
5	*Plectranthias barroi* Fourmanoir, 1982	n.i.
6	*Plectranthias bauchotae* Randall, 1980	n.i.
7	*Plectranthias bilaticlavia* Paulin & Roberts, 1987	n.i.
8	*Plectranthias chungchowensis* Shen & Lin, 1984	n.i.
9	*Plectranthias exsul* Heemstra & Anderson, 1983	n.i.
10	*Plectranthias fijiensis* Raj & Seeto, 1983	n.i.
11	*Plectranthias foresti* Fourmanoir, 1977	n.i.
12	*Plectranthias fourmanoiri* Regan, 1908	113
13	*Plectranthias gardineri* Randall, 1980	n.i.
14	*Plectranthias garrupellus* Robins & Starck, 1961	n.i.
15	*Plectranthias helenae* Randall, 1980	n.i.
16	*Plectranthias inermis* Randall, 1980	109
17	*Plectranthias intermedius* (Kotthaus, 1973)	n.i.
18	*Plectranthias japonicus* (Steindachner, 1883)	n.i.
19	*Plectranthias jothyi* Randall, 1996	n.i.
20	*Plectranthias kamii* Randall, 1980	113
21	*Plectranthias kelloggi* (Jordan & Evermann, 1903)	n.i.
22	*Plectranthias knappi* Randall, 1996	n.i.
23	*Plectranthias lasti* Randall & Hoese, 1995	n.i.
24	*Plectranthias longimanus* (Weber, 1913)	112
25	*Plectranthias maculicauda* (Regan, 1914)	113
26	*Plectranthias maugei* Randall, 1980	n.i.
27	*Plectranthias megalepis* (Günther, 1880)	n.i.
28	*Plectranthias megalophthalmus* Fourmanoir & Randall, 1979	n.i.
29	*Plectranthias morgansi* (Smith, 1961)	113
30	*Plectranthias nanus* Randall, 1980	112
31	*Plectranthias pallidus* Randall & Hoese, 1995	n.i.
32	*Plectranthias parini* Anderson & Randall, 1991	n.i.
33	*Plectranthias pelicieri* Randall & Shimizu, 1994	110
34	*Plectranthias randalli* Fourmanoir & Rivaton, 1980	n.i.
35	*Plectranthias retrofasciatus* Fourmanoir & Randall, 1979	n.i.
36	*Plectranthias robertsi* Randall & Hoese, 1995	n.i.
37	*Plectranthias rubrifasciatus* Fourmanoir & Randall, 1979	n.i.
38	*Plectranthias sagamiensis* (Katayama, 1964)	111
39	*Plectranthias sheni* Chen & Shao, 2002	n.i.
40	*Plectranthias taylori* Randall, 1980	n.i.
41	*Plectranthias vexillarius* Randall, 1980	n.i.
42	*Plectranthias wheeleri* Randall, 1980	n.i.
43	*Plectranthias whiteheadi* Randall, 1980	n.i.
44	*Plectranthias winniensis* (Tyler, 1966)	112
45	*Plectranthias yamakawai* Yoshino, 1972	n.i.

n.i. = not included

Red-blotched Perchlet *Plectranthias inermis*

Plectranthias inermis Randall, 1980. Philippines.

Indo-West Pacific, widespread in Western Pacific and reported from a few locations in the Indian Ocean, west to Mauritius. Deep coastal and inner reef drop-offs, usually at depths of over 30 m, but reported from a depth range of 15–65 m. It has been photographed on a number of occasions. A beautifully marked species, but looks more grey in deep water. Best identified by the white membrane of the 3rd spine of the dorsal fin and its blotched pattern on the body. A small species, reaching about 40 mm in total length.

A

P. inermis. Bali, Indonesia. Depth 52 m. Length 40 mm. T. Tonozuka.

B

C

P. inermis. PNG. Depth 20 m. Length 40 mm. Roger Steene.

Crested Perchlet *Plectranthias altipinnatus*

Plectranthias altipinnatus Katayama & Masuda, 1980. Izu Peninsula, Japan.

Known only from subtropical Japan, the Izu Peninsula to Okinawa region. Rocky reefs at depths of 40+ m. Small reddish species easily overlooked as colour is neutralised in deep water. Anterior dorsal-fin spines distinctive, being white and the third very tall with a flag-like flap at the tip. Orange-red dorsally and sides with red or yellow spots. Length to 55 mm.

Remarks: there are two forms, perhaps representing different species. **A** is thought to represent the true species and **B** & **C** new to science.

A

P. altipinnatus. Kochi Pref., Japan. D. 42 m. L. 40 mm. Tomonori Hirata.

B

C

P. cf *altipinnatus*. Kumejima, Japan. Depth 55 m. Length 35–40 mm. Tsuyoshi Kawamoto.

A

P. pelicieri. Mauritius. Depth 40 m. Length ~9 cm. Jan Post.

Pelicier's Perchlet
Plectranthias pelicieri

Plectranthias pelicieri Randall & Shimizu, 1994. Mauritius.

Known from Mauritius, Indian Ocean, and a small population in southern Japan, from the Okinawa to Izu region. Solitary species, usually seen in certain sections of deep rocky reefs at 33 m or more. Orange-red with yellow markings. Occurs at depths 35+ m. Length to 10 cm.

Remarks: its geographically disjuncted Indian and Pacific Ocean distribution suggests that there are two separate species, but their colouration differs only slightly. I.O form has red tips on first two dorsal fin spines.

B

C

P. pelicieri. Kumejima, Japan. Depth 35 m. Length ~60 mm. K. Yunokawa.

D

P. pelicieri. Kashiwajima, Japan. Depth 35 m. Length ~9 cm. Yasuaki Miyamoto.

P. pelicieri. Mauritius. Depth 40 m. Length ~8–9 cm. Daniel Pelicier.

Thread-tail Perchlet
Plectranthias sagamiensis

Plectranthias sagamiensis Katayama, 1964. Japan.

Pacific coast of southern Japan and reported from Indonesia. Deep rocky reefs to well over 100 m depth. Body pale with large orange blotches that may fade over most of the body in large adults. Caudal fin develops several filamentous extensions on rays, some on upper ones quite long, about snout length. Length to 10 cm.

P. sagamiensis. Izu Peninsula, Japan. Length ~7 cm. Satoshi Yamamoto.

P. sagamiensis. Osezaki, Japan. Depth 65 m. Length 9 cm. A. Mishiku.

Longfin Perchlet *Plectranthias longimanus*

Pteranthias longimanus Weber, 1913. East Indies.

Widespread Indo-West Pacific. Usually common on coastal sand slopes, but secretive amongst small boulders or isolated rock or coral outcrops and bommies with rich invertebrate and algae growth. Shallow in protected bays, and ranging to at least 75 m depth. Distinguished from similar species by the thin line, angled back from below eye and 3 white spots on the posterior end, as shown in **B**, that are always conspicuous, regardless of variation elsewhere on the body that may show additional white spots. Length to 35 mm.

A

P. longimanus. Milne Bay, PNG. Depth 15 m. Length 30 mm.

B

P. longimanus. Bali, Indonesia. Depth 6 m. Length 35 mm.

C

P. longimanus. Kochi Pref., Japan. D. 35 m. T. Hirata.

Pixy Perchlet *Plectranthias nanus*

Plectranthias nanus Randall, 1980. Guam.

Primarily known from oceanic locations of the Pacific, but also known from Japan, and Christmas Island in the eastern Indian Ocean. Very secretive in the coral sub-base rubble from shallow reef flats to almost 60 m depth. Has distinct white saddle-spot and two pupil-sized black spots on caudal peduncle. Length to 45 mm.

P. nanus. Christmas I., Indian Ocean. L. 35 mm. Jerry Allen.

Red-blotch Perchlet *Plectranthias winniensis*

Pteranthias winniensis Tyler, 1966. Seychelles.

Western Indian Ocean, Mauritius and Seychelles, south to Natal, and Red Sea, but also reported as this species from elsewhere in the Indo-Pacific. Very secretive on deep reefs, usually over 25 m depth. Brightly coloured with yellow and red, and a prominent white saddle spot at posterior end of dorsal fin base. Length to 45 mm.

A

P. winniensis. Kumejima, Japan. D. 55 m. Tsuyoshi Kawamoto.

B

P. winniensis. Aliwal Shoal, Natal, S. Africa. L. 35 mm.

Fourmanoir's Perchlet *Plectranthias fourmanoiri*

Plectranthias fourmanoiri Randall, 1980. Marshall Is.

Widespread in Western Pacific, eastern Indian Ocean and Central Pacific. Inshore lagoons to outer reefs. Depth range from 5–45 m. Best identified by the black spots at the end of both dorsal and anal fin bases, partly onto body and several broad dark bands posteriorly on the body. Length to 45 mm.

P. fourmanoiri. Christmas Island, I.O. L. 40 mm. Jerry Allen.

Morgan's Perchlet *Plectranthias morgansi*

Pelontrus morgansi Smith, 1961. Kenya.

Known only from off South Africa and Kenya. A deep water species, with 70 m the shallowest depth recorded. Pinkish above with red and yellow markings, a red band from centre of spinous part of dorsal fin running down over 9th to 16th lateral line scales grading to yellow ventrally, and white below. Length to 65 mm.

P. morgansi. Kenya. Length 60 mm. Phil Heemstra.

Kami's Perchlet *Plectranthias kamii*

Plectranthias kamii Randall, 1980. Okinawa, Japan.

Guam, Micronesia and southern Japan from Izu south to Ryukyu Islands. A deep water species, occasionally caught on line from 150–300 m depth. Colour pink to red with yellow spots and scribbles. Japanese fish are more banded and have a longer third dorsal fin spine. A large species that may reach 25 cm in length.

P. kamii. Guam. D. 150–180 m. L. 19 cm. Robert Myers.

Red-barred Perchlet *Plectranthias azumanus*

Pseudanthias azumanus Jordan & Richardson, 1910. Japan.

Izu region and further south in Japan. Occurs on deep rocky reefs at depths of 80 m or more. Mostly orange, a broad red band below end of spinous section of dorsal fin, a red blotch on caudal peduncle and a small red spot high on caudal fin base. Length to 14 cm.

Remarks: Hawaiian *Plectranthias kelloggi* shares red caudal spots, but body colouration is markedy different, and length to 21 cm.

P. azumanus. Tokai aquarium, Japan. Length 12 cm.

Red-spot Perchlet *Plectranthias maculicauda*

Serranops maculicauda Regan, 1914. Cape North, New Zealand.

Tasman Sea region of Australia and New Zealand, but two forms and Australian population may represent an undescribed species. Deep offshore reefs at 100–155 m depth. Body reddish pink with yellow shading. Large red blotch, larger than eye, on caudal peduncle and a conspicuous red spot or streak high on caudal fin base. Australian sibling lacks the distinct banding shown in New Zealand population. Length to 14 cm.

A

P. maculicauda. Bay of Islands, New Zealand.
Line, 100 m. Length ~14 cm. Craig Worthington.

B

P. cf *maculicauda*. NSW, Australia. D. 150 m. L. 12 cm. Ken Graham.

Subfamily Anthiinae, Genus *Othos* Castelnau, 1875

Masculine. Type species: *Othos cephalotes* Castelnau, 1875. (= *O. dentex*). A temperate, monotypic genus from southern Australian waters. One of the largest of the anthiids. Easy to keep, but only suitable for very large aquariums.

Harlequin Fish *Othos dentex*

Plectropoma dentex Cuvier, 1828.
King Georges sound, Western Australia.

South-western Australia and South Australia. In Victoria only known from old records, perhaps its young expatriating to the area intermittently or lived there when temperatures were higher. Occurs on shallow rocky reefs with heavy algae growth or deeper in sponge areas. Usually in caves and shows no fear of divers, often swimming right up to the camera. Variable in colour from red to green with blue spots and a series of large yellow to green blotches along lower sides. A very large anthiid, reaching 75 cm.

Remarks: superficially resembles coral trout, a reputed good table fish, but the Harlequin fish not considered to be very palatable.

O. dentex. **A** Cleaner clingfish above eyes. .**B** several cleaner clingfishes on the back. SA, southern Australia. Length 60 cm.

O. dentex. Large adult, WA, southern Australia. Depth 6 m. Length 70 cm.

Subfamily Anthiinae, Genus *Hypoplectrodes* Castelnau, 1875

Masculine. Type species: *Plectropoma nigrorubrum* Cuvier, 1828. A subtropical to temperate genus from the southern Pacific, comprising 10 species, 2 of which are undescribed. They are usually found on rocky reefs in crevices, but some sit on exposed surfaces of sponges. Few live in deep water (100+ m) and are rarely seen. Small to medium sized fishes, ranging from 10–30 cm in adult size. They feed on a variety of small swimming or crawling invertebrates or fishes. All members are easy to keep and ideal for the cold-water reef aquarium, but some grow to a moderate size.

Banded Seaperch *Hypoplectrodes nigroruber*

Plectropoma nigrorubrum Cuvier, 1828.
King Georges Sound, WA, Western Australia.

Widespread along the coast of southern Australia between Sydney and Perth, including northern Tasmania. Inhabits rocky and sponge reefs. In protected coastal bays it occurs at depths of about 10 m, but deeper on exposed reefs. A solitary species. Adults sit openly on top of boulders or sponges. Juveniles secretive in crevices. Highly variable in colour, usually body pale and 4 dark bands on posterior half of the body. Bands usually black, but sometimes red in deep water and dusky to bright red dorsally on head and anteriorly on body. Juveniles with distinctive white line on top of head from tip of snout to dorsal-fin origin. A large species, length to 30 cm.

Remarks: easily kept, but suitable only for the large home aquarium.

H. nigroruber. Jervis Bay, NSW, Australia. Depth 10 m. Length 25 cm.

H. nigroruber. Sydney Harbour, NSW, Australia. Depth 10–15 m. **B** juvenile, length 45 mm. **C** large adult with cleaner wrasse.

H. nigroruber. **D** Wilsons Promontory, Victoria. Depth 25 m. **E** Busselton, Western Australia. Depth 7 m. Length 22 cm.

H. semicinctus. Juan Fernãndez Islands, Chile. Álvaro Sepúlveda.

Chile's Half-banded Seaperch
Hypoplectrodes semicinctus

Plectropoma semicinctum Valenciennes, 1833. Juan Fernández Islands.

Juan Fernández Islands, where common, rare in the Galápagos Islands and a single record from Eastern Island. Rocky reefs, boulder habitats and large crevices at depths to about 20 m. A distinct species (in its geographical range) due to its shape and colour pattern of red to black bars. Pectoral and ventral fins yellow. Length to about 20 cm.

Remarks: de Buen, 1959 described *Scopularia rubra*, but type seems to be lost. Most authors agree that this species is synonymous.

A

H. maccullochi. Bermagui, NSW, Australia. Depth 20 m. Length 20 cm.

Australian Half-banded Seaperch
Hypoplectrodes maccullochi

Ellerkeldia maccullochi Whitley, 1928. Maroubra, Sydney, Australia.

East coast of Australia from northern New South Wales to eastern Tasmania. A common species on sponge reefs in rocky open estuaries, coastal bays and to moderate depths on off shore reefs, to at least 50 m. Often found in small loose groups and large sponges may support several individuals. Identified by the banded pattern that is usually darkest above the lateral line. Length to 20 cm.

B

C

H. maccullochi. **B** Bicheno, Tasmania. Depth 25 m. Lengths 4–5 cm. **C** northern NSW, Australia. Depth 20 m. Length 12 cm.

H. sp. Off NSW, Australia. Length 32 cm. Ken Graham.

Broad-banded Seaperch
Hypoplectrodes sp

Undescribed species.

Known only from off the NSW coast, Australia. A deep water species only caught on line from rocky reefs offshore. Deep-bodied with broad brown bands. Pectoral and ventral fins pink to orange and median fins dusky-yellow. Largest species in genus that can reach 40 cm.

New Zealand Half-banded Seaperch
Hypoplectrodes cf *maccullochi*

An undescribed species.

Known only from the Kermadec- and Three Kings Islands to East Cape, northern New Zealand. Mainly found in the offshore islands on rocky reefs, where amongst boulders or on ledges in the shallows to depths of at least 50 m. Similar to the closely related Australian Half-banded Seaperch, but banding on body is more defined, including the parts below the lateral-line, and usually black on caudal peduncle. Length to 15 cm.

A

H. cf *maccullochi.* Three King Is., New Zealand. Malcolm Francis.

B

C

H. cf *maccullochi.* Poor Knights, New Zealand. Malcolm Francis.

Red-banded Seaperch
Hypoplectrodes huntii

Plectropoma huntii Hector, 1875.
Chatman Is., off New Zealand.

Widespread New Zealand from the Three Kings Islands in the north to the Snares Islands in the south, and Chatman Islands. Occurs on coastal reefs with caves and boulders at various depths, reported to about 100 m. Identified by the distinctive banding on the body, the bands reaching ventrally and usually darkened mid-laterally. Length to about 20 cm.

A

H. huntii. Fiordland, southern New Zealand. Malcolm Francis.

B

C

H. huntii. Three King Is., northern New Zealand. Malcolm Francis.

A

H. cardinalis. Recherche Archipelago, Western Australia. Depth 15 m. L. 10 cm.

Red Seaperch
Hypoplectrodes cardinalis

Hypoplectrodes cardinalis Allen & Randall, 1990. Replacement name (pre-occupied) for *Ellerkeldia rubra* Allen, 1976. Houtman Abrolhos, Western Australia.

South-western Australia, Recherche Archipelago to Sharks Bay. Secretive on rocky reefs in 10–35 m depth, adults in the back of dark caves or under far-reaching ledges. Overall reddish pink with a broad dusky-red lateral band. Shows a pattern of pale blotches at night. A small species, length to about 10 cm.

B

C

H. cardinalis. Recherche Archipelago, Western Australia. **B** night. Depth 20 m. L. 8 cm. **C** depth 25 m. Length 9 cm.

Spotty Seaperch
Hypoplectrodes wilsoni

Ellerkeldia wilsoni Cuvier, 1828.
Geograph Bay, WA, Western Australia.

South-western Australia, Recherche Archipelago to Shark Bay. Secretive on rocky reefs in 5–35 m depth, adults in the back of caves or under far-reaching ledges. Usually only seen at night in the back of caves. A solid-bodied species with broad brown band, distinct from the generally mottled pattern all over. Length to 20 cm.

A

B

H. wilsoni. Recherche Archipelago, Western Australia. Depth 30 m. Length 18 cm.

Jameson's Seaperch
Hypoplectrodes jamesoni

Hypoplectrodes jamesoni Ogilby, 1908.
Moreton Bay, Qld, Australia.

East coast of Australia from southern Queensland to central New South Wales. Usually in rocky estuaries with sponges at a depth range of 10–30 m, but also shallower under piers that often support sponge growth on the pylons in tropical waters. A somewhat secretive species, not often noticed. Common in southern Queensland, but uncommon in the Sydney region, its southern limit. A somewhat drab species with irregular dark brown-grey blotches, sometimes forming irregular bands over the back, and a dark stripe from eye to tip of snout. Length to 12 cm.

A

H. jamesoni. Moreton Bay, Qld, Australia. Depth 5 m. Length 10 cm.

B

C

H. jamesoni. **B** Keppel Island. D. 12 m. L. 8 cm. **C** Sydney Harbour, NSW, Australia. Depth 7 m. Length 9 cm.

Black-banded Seaperch
Hypoplectrodes annulatus

Plectropoma annulatum Gunther, 1859.
Port Jackson (Sydney), Australia.

East coast of Australia from central New South Wales to Wilsons Promontory, Victoria. Protected coastal reefs and rocky estuaries. Usually seen upside-down on ceiling of caves or crevices at depths over 10 m. A stocky fish when adult, creamy with dark-brown to black bands. Large adults yellow with black bands. Length to 30 cm.

A

H. annulatus. Botany Bay, NSW, Australia. Depth 15 m. Length 18 cm.

B

C

H. annulatus. Australia. **B** Sydney Harbour. Depth 10 m. L. 6 cm. **C** Wilson's Promontory, Victoria. Depth 22 m. Typical habitat.

Subfamily Anthiinae, **Genus** *Acanthistius* Gill, 1862

Masculine. Type species: *Plectropoma serratum* Cuvier, 1828. A temperate, southern hemisphere Pacific and Atlantic genus with 8 species. Inhabits rocky reefs, usually sheltering in crevices or caves. Some species live in deep water (100+ m) and are rarely seen. Medium sized fishes, ranging from about 30–50 cm max in size. They feed on a variety of small swimming or crawling invertebrates and fishes. All members are easy to keep and ideal for the large cold-water reef aquarium and best suited for public aquariums due to their large size. In Australia these fishes are regarded as unpalatable, having very coarse flesh and little flavour.

A

Western Wirrah *Acanthistius serratus*

Plectropoma serratum Cuvier, 1828.
Western Australia.

South-western Australia, from Great Australian Bight to just north of Perth (Kalbarri). Oceanic reefs, usually secretive in caves. Small juveniles inshore, often in rockpools. Variable grey to brown with numerous dusky spots, sometimes with large dusky bands and blotches. Length to 50 cm.

B

C

A. serratus. **A** & **B** aquarium, Geograph Bay. **C** Recherche Archipelago, south-western WA, Australia. Depth 25 m. Length 35 cm.

Leopard Wirrah *Acanthistius pardalotus*

Acanthistius pardalotus Hutchins, 1981.
Western Australia.

Western Australia, ranging from Cape Leeuwin in the south to Coral and Shark Bay in the north. Occurs on exposed reefs at 10 to 25 m depth. Identified by the leopard-like spotted pattern and often several stripes radiating from behind eye. Length to 50 cm.

A. pardalotus. Shark Bay, Western Australia. Depth 20 m. Barry Hutchins.

Eastern Wirrah *Acanthistius ocellatus*

Plectropoma ocellatum Günther, 1859.
Australian seas.

South-east coast of Australia from southern Queensland to eastern Bass Strait region, including north-eastern Tasmania. A solitary species in coastal rocky bays and estuaries as well as on sheltered deep rocky reefs to at least 100 m depth. Brownish or greenish yellow with numerous blue spots on head and body, spots reducing in size proportionally with growth and become more numerous. Fins plain, bluish, darkening with age and pale blue margins. Length to 45 cm.

A

A. ocellatus. Sydney, NSW, Australia. Depth 10 m. Length 25 cm.

B

C

A. ocellatus. Juveniles. Sydney Harbour, NSW, Australia. Depth 5 m. Length 35–45 mm.

D

A. ocellatus. Montague I., NSW, Australia. Depth 10 m. Length 42 cm.

A

A. cinctus. Kermadec Is., New Zealand. Malcolm Francis.

Banded Wirrah
Acanthistius cinctus

Plectropoma cinctum Günther, 1859.
Norfolk Island.

Mainly known from the Norfolk- and Lord Howe Island regions, Kermadec Islands, and ranging south to the offshore islands of northern New Zealand. Usually in caves or below large overhangs of reef. Primarily a solitary species, but occasionally in pairs or in small aggregations where common. A distinctive species, yellowish with broad black bands on the body and black stripe from tip of snout running over eye onto first band. Length reported to 60 cm, but it is usually much smaller (40–45 cm).

B

A. cinctus. Lord Howe Island, NSW, Australia. Malcolm Francis.

Striated Wirrah
Acanthistius paxtoni

Acanthistius paxtoni
Hutchins & Kuiter, 1892.
Seal Rocks, NSW, Australia.

Known only from a few specimens from the Sydney region of the east coast of Australia. Depth range (known) 20–64 m, but is likely to be a deep-water species that lives in caves and rarely captured. May enter fish-traps. Grey with irregular dusky band and orange scribbles on body and head. Fins dusky. Largest specimen measures 30 cm.

A. paxton. Holotype. Seal Rocks, NSW, Australia. Trap 64 m. Length 30 cm.

Patagonian Wirrah
Acanthistius patachonicus

Plectropoma patachonica Jenyns, 1840. Patagonian coast.

Cool waters of Brazil, Rio De Janeiro, to northern Argentinian coast. Rocky reefs from about 20 to 70 m depth. Mostly grey to dusky with dark spots and scribbles, sometimes forming indistinct banding, a moustache-like dusky band on snout and usually a similar band from below eye to pectoral-fin base. Length to 50 cm.

A. patachonicus. Adult. Brazil. Alfredo Carvalho.

Brazilian Wirrah *Acanthistius brasilianus*

Plectropoma brasilianum Cuvier, 1828. Brazil.

Cool waters of Brazil, coastal regions of São Paulo and Rio De Janeiro. Rocky reefs from about 15 to 60 m depth. Variable from grey with thin dusky bands, to yellowish brown to dusky with pale bands. Length to 40 cm, usually 35 cm.

Remarks: local English name 'Barred Rockfish'.

A. brasilianus. Brazil. Alfredo Carvalho.

A. brasilianus. Brazil. Alfredo Carvalho.

Peruvian Wirrah *Acanthistius pictus*

Plectropoma pictum Tschudi, 1846. Peru.

Eastern Pacific, Peru and northern Chile. Deep coastal reefs. Little known species, other than being caught on line from moderate depths and their appearance in fish markets. Colour reddish after capture with a pattern of scribbled brown to orange lines and dark fins. Gill-margin orange. Length to 50 cm.

A. pictus. Northern Chile. Alvaro Sepubreda.

A. pictus. Lima, Peru. Alfredo Carvalho.

Subfamily Anthiinae, **Genus** *Epinephelides* Ogilby, 1899

Masculine. Type species: *Epinephelides leai* Ogilby, 1899 (=*Serratus armatus* Castelnau, 1875). A monotypic genus from southwestern Australia, closely related to *Hypoplectrodes*, but also has similarities to *Caesioperca.* A solitary species living in caves, but is inquisitive with divers, often resting on the bottom nearby to observe the strange invader.

A

E. armatus. Geograph Bay, Western Australia. Depth 5 m. Length 20 cm.

Breaksea Cod *Epinephelides armatus*

Serratus armatus Castelnau, 1875.
Swan R., Western Australia.

Southwestern Australia from the Recherche Archipelago to Shark Bay. Lives on rocky reefs in crevices or caves, moderately common in shallow exposed oceanic localities and often observed, as it comes out to look at divers and is easily approached at close range. Also commonly caught on line. Variable yellowish or greenish grey. Head mostly dark when small and white stripe from snout to dorsal origin. Black fades with growth to a dusky band over eye. Iris red. Length to 55 cm.

B

C

E. armatus. Western Australia. **B** Fremantle. Depth 5 m. Length 20 cm. **C** depth 10 m. Length 35 cm.

D

E. armatus. Canal Rocks, Western Australia. Depth 10 m. Length 35 cm.

SUBFAMILY ANTHIINAE, **GENUS** *Trachypoma* Gill, 1862

Neutral. Type species: *Trachypoma macracanthus* Günther, 1859. A subtropical genus that was thought to be monotypic, but 2 more species in South Africa appear to belong to this genus, one of which was only recently discovered. Similar to *Acanthistius*, but unlike members of the latter that usually swim or hover near the bottom, the species in *Trachypoma* rest on the substrate. They have rounded fins, small anal fin and long pectoral fins, latter used to wedge themselves against rock-faces in swells or strong currents

Strawberry Cod *Trachypoma macracanthum*

Trachypoma macracanthus Günther, 1859.
Norfolk Is.

Known from the Norfolk, Lord Howe Islands, Kermadecs, northern New Zealand and NSW, Australia. Sheltered coastal reefs, usually in boulder habitats or narrow crevices. At all sizes orange with numerous tiny white spots all over, including fins, but sometimes shows a pattern of pale blotches, especially at night. Length to 40 cm, but usually less than 30 cm.

A

T. macracanthum. Sydney, NSW, Australia. Depth 10 m. Length 20 cm.

B

C

T. macracanthum. Sydney, NSW, Australia. **B** aquarium. Length 10 cm. **C** night. Depth 10 m. Length 20 cm.

Aliwal Koester *Trachypoma* sp

Undetermined species

Eastern South Africa, from Port Alfred to the Aliwal Shoal. Rocky reefs, secretively in caves or crevices, in depths to at least 40 m. Rests on sides of rocks, wedges itself with its pectoral fins and probably well adapted to hold on in strong currents or surge. Identified by its shape and mottled colouration with a series of pale blotches. Length probably to 30 cm.

Remarks: similar to *Trachypoma sebastoides* (Castelnau, 1861), usually placed in *Acanthistius*).

A

B

T. sp. Aliwal Shoal, South Africa. **A** from 25 m. **B** depth 25 m. Length 20 cm.

SUBFAMILY EPINEPHELINAE, **GENUS** *Alphestes* Bloch & Schneider, 1801

Masculine. Type-species: *Epinephelus afer* Bloch, 1793. A small genus, thought to comprise 3 species, 2 in the eastern Pacific and one in the western Atlantic. These fishes are similar to some of the small rock-cod species (*Epinephelus*) in the Indo-Pacific, but have larger eyes, shorter snout, and a large single spine (usually hidden by skin) on the lower posterior corner of their operculum, directed down and forward, versus 3 spines, directed backward. The species, *Alphestes multiguttatum,* which is not included, is grey with numerous small dusky spots whilst the pectoral fins are barred in juveniles and spotted in adults.

1 *Alphestes afer* (Bloch, 1793)	Guinea (w. Atlantic?)
2 *Alphestes immaculatus* Breder, 1936	Gulf of California, Mexico
3 *Alphestes multiguttatum* (Günther,1867)	Panama

Atlantic Mutton Hamlet
Alphestes afer

Epinephelus afer Bloch, 1793. Guinea.

Widespread tropical to subtropical western Atlantic, Florida to Argentina, and recently a specimen from a fish market in São Tomé (eastern Atlantic) was identified by Phil Heemstra as this species. Occurs on rocky reefs and in algae habitat from very shallow to about 50 m depth. Colour variable, usually similar to dominant algae colour in the living space. Mainly red in deep water. Length to about 35 cm.

A. afer. Brazil. Length 30 cm. Alfredo Carvalho.

Pacific Mutton Hamlet *Alphestes immaculatus*

Alphestes immaculatus Breder, 1936.
Gulf of California, Mexico.
Alphestes galapagensis Fowler, 1944. Galápagos Islands.

Eastern Pacific: Gulf of California to Peru and offshore islands, including Galápagos. West Atlantic: Rocky reefs, usually resting on the bottom during the day and hunting from dusk till dawn. Juveniles reddish brown with white spots, which change to an irregular blotched pattern with growth. Median fins with small pale or white spots. Pectoral fins barred to spotted from juvenile to adult respectively. Length to 33 cm, but usually much smaller.

A. immaculatus. Galápagos I. **A** juvenile. Length ~ 10 cm. **B** adult. Length ~25 cm. Paul Humann.

Subfamily Epinephelinae, **Genus** *Paranthias* Guichenot, 1868

Masculine. Type-species: *Serranus furcifer* Valenciennes, 1828. Two very similar species are recognised, divided between eastern Pacific and Atlantic, both with a very large geographical distribution. Superficially this genus resembles the subfamily Anthiinae, but is more closely related to the rock-cods and groupers, subfamily Epinephelinae. However, like the anthiids, they are planktivores and may form large schools high above the bottom when feeding. Their shape and large forked caudal fin is typical amongst planktivorous reef-fishes (eg. anthiids, Caesionidae and planktivorous lutjanids).

These fishes appear to be of little interest to aquarists and are rarely caught on line.

Paranthias furcifer. Type-species for the genus. Buraco do Inferno, Brazil. Depth 10 m. Length 15 cm. João Gasparini.

Paranthias colonus. Massing midwater in a large school to feed on zooplankton. Galápagos Islands. Jerry Allen.

A

P. furcifer. Salvador, Brazil. Length about 20 cm. Bertran Feitoza.

Atlantic Creolefish *Paranthias furcifer*

Serranus furcifer Valenciennes, 1828. Brazil.

Western Atlantic with a number of populations: West Indies north to Florida and Bermudas, and south to Brazil, and eastern Atlantic in the Gulf of Guinea. Occurs in small groups over rocky substrates in the eastern Atlantic, whilst in the western Atlantic they may form very large schools on reefs in oceanic localities. Mainly swim near the bottom deeper than 20 m, known as deep as 70 m, but occasionally may follow food up to near the surface. In most places they grow to about 20 cm in length, rarely 25 cm, but individuals of the population in Brazilian can reach 40 cm.

Remarks: 2 different species may be involved.

B

P. furcifer. Night. St. Vincent, L. Antilles. Roger Steene.

C

D

P. furcifer. **C** night. Bonaire. Depth 12 m. **D** Guadeloupe, Lesser Antilles. Depth 22 m. Length 20 cm. Patrick Louisy.

E

P. furcifer. Bonaire, Netherlands Antilles. Depth 12 m. Length 18 cm. Roger Steene.

Pacific Creolefish *Paranthias colonus*

Serranus colonus Valenciennes, 1846. Galápagos Is.

Widespread eastern Pacific from Galápagos Islands and Peru in the south, to Gulf of California and offshore in the Cocos and Clipperton Islands in the north. Forms massive schools in some places such as the Galápagos Islands. Adults in the Galápagos greenish grey with a reddish tail, usually showing a series of white spots on back when on the bottom. Clipperton adult fish look somewhat different, being pale-blue with red tail. Juveniles yellowish with a few dark spots on the back. Length to 25 cm

A

P. colonus. Large adult. Galápagos Islands. Jerry Allen.

B

C

P. colonus. Clipperton I., Eastern Pacific. Jerry Allen.

D

P. colonus. Juvenile. Galápagos Is. Jerry Allen.

E

P. colonus. Sea of Cortes, Mexico. D. 15 m. L. 20 cm. Roger Steene.

F

P. colonus. Sea of Cortes, Mexico. Depth 15 m. Length 22 cm. Roger Steene.

Subfamily Liopropomatinae, Genus *Liopropoma* Gill, 1861

Neutral. Type species: *Perca aberrans* Poey, 1860. A genus of small fishes, comprising 18 tropical species, divided between Indo-Pacific and western Atlantic Oceans. Secretive fishes, many of which only recently discovered, featuring a dorsal fin with a very deep notch over the posterior half of the spinous section, dividing it into 2 fins in most species with short embedded spines in between. Their shape mirrors the ventral and anal fins. Also has large scales on the snout and the area below the nostrils naked. The genus *Pikea* is closely related and can be differentiated by the notched, but continuous dorsal fin, having small scales covering all of the snout, and growing to a much larger size.

Being small and often colourful, these fishes are in high demand in the ornamental fish trade. They are small predators that feed on a variety of small crawling or swimming invertebrates and small fishes. This makes them ideal for reef aquariums, and their secretive nature makes them more interesting as you get an occasional glimpse, rather than them being in full view at all times.

1	*Liopropoma aberrans* (Poey, 1860)	n.i.
2	*Liopromoma africanum* (Smith, 1954)	132
3	*Liopromoma carmabi* (Randall, 1963)	131
4	*Liopromoma collettei* Randall & Taylor, 1988	131
5	*Liopromoma danae* (Kotthaus, 1970)	n.i.
6	*Liopromoma dorsoluteum* Kon, Yoshino & Sakurai, 1999	n.i.
7	*Liopromoma eukrines* (Starck & Courtenay, 1962)	133
8	*Liopromoma flavidum* Randall & Taylor, 1988	n.i.
9	*Liopromoma incomptum* Randall & Taylor, 1988	n.i.
10	*Liopromoma longilepis* Garman, 1899	n.i.
11	*Liopromoma mitratum* Lubbock & Randall, 1978	133
12	*Liopromoma mowbrayi* Woods & Kanazawa, 1951	132
13	*Liopromoma multilineatum* Randall & Taylor, 1988	133
14	*Liopromoma pallidum* (Fowler, 1938)	n.i.
15	*Liopromoma rubre* Poey, 1861	130
16	*Liopromoma susumi* (Jordan & Seale, 1906	131
17	*Liopromoma swalesi* (Fowler & Bean, 1930)	131
18	*Liopromoma tonstrinum* Randall & Taylor, 1988	132

n.i. = not included

L. rubre. Aquarium. Length 75 mm. Scott Michael.

Peppermint Cave-bass *Liopropoma rubre*

Liopropoma rubre Poey, 1861. Cuba.

Widespread in the Caribbean to Bahamas and southern Florida. Lives in dark caves and below overhangs of reefs. Moderately common, but rarely observed in the wild. Shallow to at least 50 m deep. A distinctive species with its brown to black longitudinal stripes with pale yellow stripes running through the middle of the pale spaces between. Second dorsal, anal and caudal fin tips with black blotches and pale-blue margins. Length to 9 cm.

L. rubre. Cayman I. Length 8 cm. Paul Humann.

L. rubre. Aquarium. Origin unknown. Length 65 mm.

L. sp. Principe, eastern Atlantic. Length about 10 cm. Peter Wirtz.

Principe Cave-bass *Liopropoma* sp

Undetermined species.

Known only from the photograph included here, taken a Principe Island, off the Equatorial Guinea coast, just north of the Equator in the eastern Atlantic. Depth about 40 m. Length abou 10 cm.

Candy Cave-bass *Liopropoma carmabi*

Chorististium carmabi Randall, 1963. Puerto Rico.

Widespread in the West Indies, ranging to Florida Keys in the north and south to north-eastern Brazil. A rarely observed species that can be found as shallow as 15 m in shaded parts of reefs, but lives usually much deeper and reported to 70 m. Readily identified by its gaudily-coloured, striped pattern. Length in West Indies to about 6 cm, but can reach 10 cm in the cooler Brazilian waters.

A *L. carmabi.* Guarapari, Brazil. D. 30 m. L. 10 cm. João L. Gasparini.

B *L. carmabi.* Aquarium. Length 60 mm. Scott Michael.

Orange-lined Reef-basslet *Liopropoma swalesi*

Chorististium swalesi Fowler & Bean, 1930. Sulawesi.

Western Pacific. Known from a few localities in Indonesia, Rabaul, New Britain and Philippines. Occurs on coastal inner-reef slopes in rich coral growth at depths of about 25 m. Identified by the orange stripes and large white-edged black ocelli on the dorsal and anal fins. A small species, length to about 50 mm.

A *L. swalesi.* Aquarium. Flores, Indonesia. From 25 m. L. 45 mm.

B *L. swalesi.* Aquarium. Length 45 mm. Scott Michael.

Collette's Reef-basslet *Liopropoma collettei*

Liopropoma collettei Randall & Taylor, 1988. Hawaiian Islands.

Western Pacific, Philippines, Indonesia and Papua New Guinea, and Hawaiian Islands. Inhabits shallow protected reefs. Secretive and rarely seen in the wild. Identified by the 3 principal black to red stripes mid-laterally along the body reaching to caudal-fin base. Length to 85 mm.

L. collettei. Aquarium. Indonesian specimens. Takeshi Aoki.

Pin-striped Reef-basslet *Liopropoma susumi*

Chorististium susumi Jordan & Seale, 1906. Western Samoa.

Widespread Indo-Pacific. Secretive in crevices and caves, especially along outer-reef walls to depths of about 30 m. Readily identified by the many thin, evenly spaced horizontal lines. Length to 92 mm.

L. susumi. Maldives. In cave. Depth 15 m. Length 85 mm.

L. tonstrinum. Christmas I., I.O. D. 35 m. L. 35 mm. Jerry Allen.

Candy Reef-basslet *Liopropoma tonstrinum*

Liopropoma tonstrinum Randall & Taylor, 1988. Palau Is.

Western-Central Pacific and Christmas Island in the eastern Indian Ocean. A rarely observed species and mainly seen on reefs when flushed out into open by using chemicals. Most specimens were collected at depths between 40 and 50 m, few shallower at 11–12 m. Identified by the pale body, parallel stripes that start as yellow on the head, changing to red along most of body and caudal fin. Length to 80 mm.

African Reef-basslet *Liopropoma africanum*

Chorististium africanum Smith, 1954. Mozambique.

Western Indian Ocean, African coast to Maldives. Secretive in dark caves or in the back of areas below large reef overhangs. Usually along outer-reef walls at depths over 10 m. Pinkish with yellow stripes on head and white longitudinal stripes along body. Length to 85 mm.

A

B

C

L. africanum. Maldives. In cave. Depth 15 m. Lengths 75–85 mm.

Pink Cave-bass *Liopropoma mowbrayi*

Liopropoma mowbrayi Woods & Kanazawa, 1951. Bermuda.

Widespread in the Caribbean and ranges north to southern Florida and Bermuda, but not so common in latter two localities. A shy species that quickly retreats to the dark recesses of caves or holes in the reef. Lives at moderate depths over about 25 m. Lacks longitudinal lines like other species in the region and is mainly plain pink to salmon coloured, but has in common the black markings with pale blue edges on the posterior tips of the median fins. Length to 9 cm.

A

B

L. mowbrayi. **A** Caribbean. Length 65 mm. Frank Schneidewind. **B** Belize. Length ~9 cm. Paul Humann.

Yellow Reef-basslet *Liopropoma multilineatum*

Liopropoma multilineatum
Randall & Taylor, 1988. Hawaiian Islands.

Western Pacific, Philippines, Indonesia to north-western Australia, and into Central Pacific to Micronesia and Fiji. Secretive in the back of large caves along steep outer-reef walls at depths from about 25 m. Readily identified by its unusual colouration with a large yellow area over most of the body and abrupt change to red and white on the posterior section, seemingly belonging to another fish, the red looking grey in the natural light. Length to 80 mm.

L. multilineatum. Togean Islands, Sulawesi. Depth 25 m. Length 80 mm.

Arabian Reef-basslet *Liopropoma mitratum*

Liopropoma mitratum Lubbock & Randall, 1978.
Saudi Arabia, Red Sea.

Appears to be widespread in the Indo-West Pacific, ranging from its Red Sea type locality to Christmas Island in the eastern Indian Ocean and many in the Pacific, if all represent this species. Rarely seen in the wild and specimens were collected on shallow reef flats to 46 m depth. Plain pinkish brown with yellow fins and some yellow markings on the head. Length to 9 cm.

L. mitratum. Aquarium. Length 85 mm. Scott Michael.

Wrasse Cave-bass *Liopropoma eukrines*

Chorististium eukrines Starck & Courtenay, 1962.
Florida, U.S.A.

Western Atlantic, northern Carolina to Florida. Deep rocky reefs in caves and crevices from about 30 to 150 m depth. Pinkish brown with mid-lateral broad, dark stripe and pale longitudinal lines above and below. Caudal fin with white posterior margin and adults with additional black submarginal stripe next to the white posterior margin. Length to 13 cm.

L. eukrines. Cape Canaveral, Florida. Paul Humann.

Subfamily Liopropomatinae, **Genus** *Pikea* Steindachner, 1874

Feminine. Type species: *Grystes lunulata* Guichenot, 1863. Comprising 10 tropical to subtropical described species plus a number of apparently undescribed ones. Northern Pacific, primarily the Japanese region to Eastern Pacific, western Indian Ocean and one in western Atlantic. Cave dwellers in rocky reefs, mostly in deep water. They have a single but usually a deeply notched dorsal fin and grow to a moderately large size, ranging from 18 to over 35 cm in length.

Predatory fishes, easily kept in captivity, but rarely collected and require a large tank with reef formations.

Suzuki's Reef-bass *Pikea lemniscata*

Liopropoma lemniscatum Randall & Taylor, 1988. Japan.

Known only from the Pacific coast of Honshu and the Ryukyu Islands, Japan. Deep reefs amongst boulders or in caves and crevices from depths of about 45 m. A mainly dark coloured species with two longitudinal white stripes and median fins with yellow streaks. Length to 18 cm.

P. lemniscata. Osezaki, Japan. Depth 50 m. Length 15 cm. A. Mishiku.

Black-stripe Reef-bass *Pikea latifasciata*

Pikea latifasciata Tanaka, 1922. Japan.

Known only from the Pacific coast of Honshu, Japan and southern Korea. Found on steep rock-walls in caves and crevices from depths of about 25 m down to at least 60 m. Bright yellow when small and a black mid-lateral stripe from tip of snout to caudal fin base. Large adults more pinkish, especially the head. Length to 20 cm.

A

P. latifasciata. Oshima, Japan. Depth 35 m. Length 16 cm.

B

C

P. latifasciata. Izu Oceanic Park, Japan. Depth 30 m. Length 20 cm.

Yellow-stripe Reef-bass
Pikea cf *latifasciata*

Undetermined species.

Known only from deep reefs off the south coast of KwaZulu-Natal, South Africa. Occasionally seen at depths over 30 m. Pinkish below and yellow above. A more or less distinct yellow band from tip of snout, running through eye, to upper caudal peduncle. Length to about 20 cm.

A

B

P. cf *latifasciata.* Adult. KwaZulu-Natal, South Africa. Depth about 40 m. Length ~20 cm. Dennis King.

Hawaiian Reef-bass
Pikea aurora

Pikea aurora Jordan & Evermann, 1903. Hawaiian Islands.

Known only from the Hawaiian Islands. Occurs on deep reefs at a depth range of about 50 to 200 m. Colour orange-pink with yellow stripes on head and along horizontal margins of median fins. Often with small dark spots scattered on body and caudal fin. Length to 20 cm.

A

P. aurora. Adult. Hawaii. Depth about 50 m. John Hoover.

B

C

P. aurora. Adult. Hawaii. Depth about 50 m. Length ~20 cm. John Hoover.

P. fasciata. **A** Galápagos I. Paul Humann. **B** aquarium. Length 10 cm. Scott Michael.

Pacific Reef-basslet *Pikea fasciata*

Liopropoma fasciatum Bussing, 1980. Off Isla del Caño, west of Osa Peninsula, Pacific coast of Costa Rica.

Eastern Pacific, southern California to Panama and in the Galápagos. Deep rocky reefs, in caves and crevices from 40 to 250 m depth. Reddish when small to dark purplish black when adult, with yellow longitudinal lines and yellow spots and scribbles on cheek. Length to 16 cm.

P. cf *aurora.* Guam. Depth 165 m. Length 20 cm. Robert Myers.

Yellow-tail Reef-bass *Pikea* cf *aurora*

Undetermined.

Known only from Guam. Deep reefs. Appears to be closely related to *Pikea aurora,* but also shows some similarity to *P. latifasciata.* Head colour is similar to *P. aurora* whilst colour and the shape of the median fins are shared with *P. latifasciata.* Length 20 cm.

P. lunulata. Guam. Depth 165 m, Robert Myers.

Réunion Reef-bass *Pikea lunulata*

Crystes lunulatus Guichenot, 1862. Réunion.

Various oceanic locations from the Western Indian to Pacific Ocean if all represent the same species. Deep water, 100+ m depth. Head pink with a distinct yellow stripe from snout through eye. Pink continuous on body and changes gradually to yellow along its back and on tail. Median fins mostly bright yellow and caudal fin is shallowly lunate. Length in Indian Ocean to 15 cm, larger in Pacific to 24 cm.

P. maculata. Guam. Depth 150–180 m. Robert Myers.

Red-spotted Reef-bass *Pikea maculata*

Pikea maculatus Döderlein, 1883. Tokyo, Japan.

North Western-Pacific, Japan and Korean region to Hawaiian Islands. Deep water, usually in almost 300 m, shallowest 100 m off Okinawa. Readily identified by the many red spots over most of the head, body and tail. A large species, length to 27 cm.

SUBFAMILY LIOPROPOMATINAE, GENUS *Rainfordia* McCulloch, 1923

Feminine. Type species: *Rainfordia opercularis* McCulloch, 1923. Monotypic genus, similar to *Liopropoma*, but much more depressed, head broadened and the two dorsal fins widely separated.

A little known species, but behaviour much like those in the genus *Liopropoma* and aquarium requirements are likely to be identical. Should be ideal for a living-reef kind of set-up.

Rainford's Reef-basslet *Rainfordia opercularis*

Rainfordia opercularis McCulloch, 1923.
Queensland, Australia.

Tropical Australian waters. Sheltered coastal and inner reefs. Typically in the back of large caves, usually near or in small holes and narrow ledges. Rarely seen due to its cryptic behaviour, but is readily identified by its slender and low profile appearance, the body height no more that the height of the caudal fin, and the colour pattern of longitudinal lines along the body and a small but distinctive ocellus on the base of the caudal fin. Length to 12.5 cm.

A

B

R. opercularis. Moreton Island, Qld, Australia. Depth 12 m. Length 11 cm. Neville Coleman.

C

R. opercularis. Moreton Island, Qld, Australia. Depth 12 m. Length 11 cm. Neville Coleman.

Family Serranidae, **Subfamily** Grammistinae

A small group of fishes, commonly known as soapfishes, often treated in their own family Grammistidae, or as serranids as such, but are now regarded as a subfamily of Serranidae. At some stage, the genus *Pseudogramma* was treated as a subfamily, Pseudogramminae, of Grammistidae. The subfamily Grammistinae comprises 10 genera, most with 1 or 2 species and confined to the Indo-Pacific, but the Atlantic and eastern Pacific genus *Rypticus* comprises 11 species, as does the Pacific and Atlantic *Pseudogramma*. With the exception of the small species, such as the *Pseudogramma*, these fishes are covered with mucus that contains grammistin toxin. It has a bitter taste and serves to deter predators and froths-up the water like soap, hence the common name applied to the group. More toxin is produces under stress and when confined in a small volume of water it may soon die and any other fish kept with it. Handling these fishes may cause a skin rash, especially in sensitive areas, as an ichthyologist found out when diving and picking up specimens after a rotenone kill, running out of space, and though using his bathers would be a good alternative!

If collecting for aquarium use, great care has to be taken when handling these fishes. When initially captured, either the water has to be replaced regularly or flushed through the space in which these fishes are kept, and it is best to keep them separate from other fishes to avoid problems. Once placed in the aquarium and taken care of properly, problems are unlikely. Some species can attain a moderate size (20–30 cm), and have a relatively large mouth and expandable stomach that can accommodate large prey and may eat smaller fishes. They are best kept in reef aquariums, and with fishes of similar size.

Genera Contents

Subfamily Grammistinae, Genus *Belonoperca* Fowler & Bean, 1930

Feminine. Type species: *Belonoperca chabanaudi* Fowler & Bean, 1930. Two species, Indo-Pacific. One commonly seen on tropical reefs, whilst *B. pylei* Baldwin & Smith, 1998, is known only from about 100 m in the Cook Islands (not included here). Slender species with a pike-like head, mouth extending to below eye when adult. The dorsal, anal and ventral fin arrangement, and body shape may cause them to be confused with apogonid fishes.

Arrow-head Soapfish *Belonoperca chabanaudi*

Belonoperca chabanaudi Fowler & Bean, 1930. Celebes.

Tropical Indo-Pacific from east Africa to Central Pacific. Typically seen in large caves and overhangs of outer reefs along steep walls. Stays in the shade, but swims about openly. May look almost black in a dark cave, but the yellow saddle-spot on the caudal peduncle stands out and is eye-catching. Generally bluish grey with numerous small dusky spots. Usually keeps fins erect and spinous dorsal and ventral fins are marked with black blotches and pale blue. Small juveniles have a leopard-like pattern of dark spots. Length to 15 cm.

A

B. chabanaudi. Maldives. Depth 30 m. Length 14 cm.

B

C

B. chabanaudi. **B** Madang, PNG. Length 25 mm. Jerry Allen. **C** Manado, Indonesia. Depth 25 m. Length 9 cm.

D

E

B. chabanaudi. **D** Madang, PNG. Length 65 mm. Jerry Allen. **E** Milne Bay, PNG. Depth 25 m. Length 12 cm.

F

B. chabanaudi. Flores, Indonesia. Depth 25 m. Length 15 cm.

SUBFAMILY GRAMMISTINAE, GENUS *Grammistes* Bloch & Schneider, 1801

Masculine. Type species: *Perca sexlineata* Thunberg, 1792. Indo-Pacific, monotypic, but has Indian and Pacific Ocean forms that differ in colour patterns. Common coastal and inner reef species in caves or below overhangs. Mainly black, but distinctly marked with white or pale yellow stripes. Body only slightly compressed and fins mostly transparent, giving it a rather thick and heavy appearance.

Easily kept in captivity, but can eat fairly large prey including other fishes.

Lined Soapfish *Grammistes sexlineatus*

Perca sexlineata Thunberg, 1792. Japan or East Indies.

Widespread Indo-West Pacific. Juveniles and subadults common, but secretive on shallow coastal reefs, usually in crevices, whilst adults are rarely seen and apparently move to deep water. Juveniles readily identified by their longitudinal striped pattern. Number of stripes increase with age and eventually break up into a series of blotches. Pattern differs between Indian and Pacific Oceans, which is especially obvious on cheek (**G** & **H**). Length to 27 cm.

A

G. sexliniatus. Adult. Length 25 cm. Sydney Aquarium

C

G. sexliniatus. Qld, Australia. Length 15 mm.

B

G. sexliniatus. Aliwal Shoal, Durban, South Africa. D. 25 m. L. 12 cm.

D

G. sexliniatus. Guam. D. 10 m. L. 65 mm.

E

F

G. sexliniatus. **E** Flores, Indonesia. Depth 10 m. Length 20 cm. **F** Mabul, Malaysia. Depth 5 m. Length 16 cm.

G

H

Grammistes sexliniatus. Indian Ocean pattern. **G** Maldives. Length 18 cm. **H** Pacific Ocean pattern.Kerama, Japan. Length 12 cm.

I

G. sexliniatus. Flores, Indonesia. Depth 25 m. Length 22 cm.

J

G. sexliniatus. Yawning. Bali, Indonesia. Depth 20 m. Length 10 cm.

SUBFAMILY GRAMMISTINAE, GENUS *Pogonoperca* Günther, 1859

Feminine. Type species: *Pogonoperca ocellata* Günther, 1859. Comprises two similar species, divided between tropical regions of the Indian and Pacific Oceans. Not often seen, but can be locally common in rocky habitats subject to strong currents. Unusually marked fishes with a peculiar dermal appendage under the mouth when adult.

Easily kept in the aquarium, but may prey on other fishes and grows to a moderate size.

A

P. punctata. Yawning. Okinawa aquarium, Japan. L. 20 cm.

B

P. punctata. Christmas I., Indian Ocean. D. 30 m. Jerry Allen.

Spotted Soapfish *Pogonoperca punctata*

Grammistes punctatus Valenciennes, 1830. Sw Pacific.

Widespread West Pacific and ranging into Indian Ocean to Christmas Island. Seen in rocky reefs in current prone habitats at depths of 20 m or more. Juveniles black with white spots. Black reduces with growth to black saddles over back, at eyes, nape, dorsal fins and caudal peduncle. Body grey to yellowish with small white spots. Fins mostly clear, 1st dorsal fin with black. Length to 34 cm.

C

P. punctata. Mabul, Malaysia. D. 5 m. L. 20 mm.

D

E

P. punctata. Sangihe Islands. Depth 20 m. Length 18–20 cm.

Snowflake Soapfish *Pogonoperca ocellata*

Pogonoperca ocellata Günther, 1859. Mauritius.

Widespread Indian Ocean, ranging east to Sumatra and southern Java. Replaced in Bali and further east and north by its Pacific sibling. Adults have white spots all over that are close-set and form reticulated patterns, including on the fins. The first dorsal fin is pale yellow in adults. Deep water coral heads and rocky reefs, usually over 25 m. Length to 27 cm.

A

P. ocellata. Maldives. D. 42 m. L. 25 cm.

B

C

D

P. ocellata. Pulau Weh, north Sumatra, Indonesia. **B** length 15 cm. **C** & **D** depth 28 m. Length 25 cm. Takamasa Tonozuka.

E

P. ocellata. Maldives. Depth 42 m. Length 25 cm.

Subfamily Grammistinae, **Genus** *Aulacocephalus* Temminck & Schlegel, 1842

Masculine. Type species: original description without species, *Aulacocephalus temminckii* added by Bleeker, 1854. Monotypic genus, widespread Indo-Pacific.

Feeds on shrimp-like crustaceans and small fishes. Easily kept, but prefers cool-tropical to subtropical and rocky habitats, and grows to a moderate size which makes them more suitable for a public aquarium.

Gold-ribbon Cod *Aulacocephalus temminckii*

Aulacocephalus temminckii Bleeker, 1854.
Nagasaki, Japan.

Widespread Indo-Pacific, but scattered localities, west to South Africa, east to Rapa, north to Japan and south to Australia and New Zealand. Rocky reefs, sometimes in small loose aggregations. Deep in tropical zones, seemingly only in areas with cool upwellings. Has been observed in coelacanth caves in the Comores to depths of 180 m (Heemstra, pers. comm.) Readily identified by the bright blue body and golden stripe running from snout, through eye, to base of caudal fin. Length to 35–40 cm, depending on locality, growing largest in cool zones.

A

B

C

A. temminckii. Izu Oceanic Park, Japan. Depth 30–35 m. Lengths 20–25 cm.

D

A. temminckii. Aliwal Shoal, Durban, South Africa. Depth 38 m. Length 30 cm.

A. temminckii. Japan. Above: Oshima, Japan. Depth 30–35 m. Lengths 25–28 cm. Below: Izu Oceanic Park. Depth 35 m. Length 20 cm.

Subfamily Grammistinae, Genus *Diploprion* Cuvier, 1828

Masculine. Type species: *Diploprion bifasciatum* Cuvier, 1828. Comprises two species, one restricted to Red Sea, and the other widespread Indo-Pacific. Inshore, coastal reefs and estuaries to deep offshore on muddy substrates. Not as secretive as other soapfishes. Deep bodied and moderately compressed, mouth large and protractible.

Easily kept in the aquarium, but may prey on other (relatively large) fishes and can grow to a moderately large size.

Yellow Emperor *Diploprion bifasciatus*

Diploprion bifasciatum Cuvier, 1828. Java.

Widespread West-Pacific and eastern Indian Ocean, ranging west to the Maldives, and to subtropical zones as larval expatriates. Mainly coastal habitats, often in silty conditions, but also in outer reef lagoons where on open substrates with remote coral-heads. Occurs singly or in small groups. Colour usually yellow with a black bar over head and a second bar from tip of spinous dorsal fin to anus, latter broadening with age, often extending over tail end of body. Occasionally body all black with yellow soft dorsal, anal and caudal fins. Juveniles appear to mimic the venomous *Meiacanthus* sabre-tooth blennies and are grey in Java. Length to 25 cm.

A

D. bifasciatus. Southern Queensland, Australia. D. 10 m. L. 22 cm.

B

D. bifasciatus. Northern Queensland, Australia. D. 15 m. L. 20 cm.

C

D. bifasciatus. NSW, Australia. D.6 m. L. 5 cm.

D

D. bifasciatus. Kochi Pref, Japan. L. 4 cm. T. Hirata.

E

F

D. bifasciatus. **E** Java, Indonesia. Depth 5 m. Length 20 cm. **F** Japan. Depth 15 m. Length 22 cm.

G *D. bifasciatus.* Keppel I, Qld, Australia. D. 6 m. L. 16 cm.

H *D. bifasciatus.* Sulawesi, Indonesia. D. 10 m. L. 18 cm.

I, J *D. bifasciatus.* Java, Indonesia. Depth 5 m. Lengths 45 & 95 mm.

Red Sea Soapfish *Diploprion drachi*

Diploprion drachi Roux-Estève, 1828. Red Sea.

Endemic to area of the Red Sea and Gulf of Aden. Solitary on coral reefs at various depths in still waters. Sneaks up on prey by swimming next to larger non-predatory fish such as herbivores (see below). Easily identified by its shape and colour, deep-bodied and mainly bluish grey with a distinct yellow patch on eye and a black band along dorsal-fin base. Length to 14 cm.

A, B *D. drachi.* Egypt, Red Sea. **A** swimming next to *Acanthurus sohal.* **B** adult. **A** & **B** Frank Schneidewind.

Subfamily Grammistinae, **Genus** *Rypticus* Cuvier, 1829

Masculine. Type species: *Anthias saponaceus* Bloch & Schneider, 1801. Comprises 10 species. The majority of these soapfishes are found in the western Atlantic and 3 are known from the eastern Pacific. They feature just 2 or 3 spines in the dorsal fin and have none in anal fin. Ventral fins small, situated anterior to or just below pectoral fin. Head long, flat dorsal profile and eyes placed dorsally above end of mouth. Usually secretive in rocky reefs, especially when juvenile, and mainly adults are occasionally observed.

These fishes have a moderately large mouth and a flexible stomach. Diet consists of fishes and invertebrates such as shrimps, often taken at dusk or at night. Small juveniles with spots or lines may have some attraction for an aquarist, but most adults are usually rather dull.

1	*Rypticus bicolor* Valenciennes, 1846	148	6	*Rypticus maculatus* Holbrook, 1855	151
2	*Rypticus bistrispinus* (Mitchill, 1818)	150	7	*Rypticus nigripinnis* Gill, 1861	150
3	*Rypticus bornoi* Beebe & Tee-Van, 1928	n.i.	8	*Rypticus randalli* Courtenay, 1967	n.i.
4	*Rypticus brachyrhinus* Courtenay, 1967	n.i.	9	*Rypticus saponaceus* (Bloch & Schneider, 1801)	149
5	*Rypticus courtenayi* McCarthy, 1979	n.i.	10	*Rypticus subbifrenatus* Gill, 1861	151

Mottled Soapfish *Rypticus bicolor*

Rypticus bicolor Valenciennes, 1846. Galápagos Is.

Eastern Pacific, Galápagos Is., Baja California to Peru and offshore islands. Rocky reefs, usually in crevices, but adults may openly move about. Dusky with pale irregular blotches and mottling. White spots more defined in juveniles and often have black centres. Length to 22 cm.

A

B

R. bicolor. Galápagos Is. **A** adult, length 16 cm. Paul Humann. **B** juvenile, 65 mm. Jerry Allen.

C

R. bicolor. Sea of Cortez. Depth 18 m. Length 22 cm. Roger Steene.

Greater Soapfish *Rypticus saponaceus*

Anthias saponaceus Bloch & Schneider, 1801. Florida, U.S.A.

Western Atlantic, Florida and West Indies to Brazil; and eastern Atlantic, Gulf of New Guinea, Cape Verde Islands and St. Helena, but colour patterns differ considerably between west and east populations. Occurs in various reef habitats and is reported between 2 and 60 m depth, but this is likely to differ in range between habitats or geographical regions. Western populations are blue-grey with pale blotches, whilst eastern populations are brown to uniformly black. Length to about 32 cm.

A

R. saponaceus. Rasas I., Brazil. D. 15 m. L. 28 cm. Sergio Floeter.

B C

R. saponaceus. Juvenile and adult. Saõ Tomé, eastern Atlantic. Peter Wirtz.

D

R. saponaceus. Abrolhos I., Brazil. Depth 15 m. Length 28 cm. Jerry Allen.

A

R. nigripinnis. Off Panama, Pacific. Length 18 cm. Jerry Allen.

Panama Soapfish
Rypticus nigripinnis

Rypticus nigripinnis Gill, 1861. Panama (Pacific).

Eastern Atlantic, mainland from Gulf of California to Peru. Secretive in rocky reefs. Small juvenile mostly black and yellow with two white lines radiating from behind eye, lower reaching ventral part of caudal-fin base. Lines shorten with growth, and body develops a spotted pattern that becomes more blotchy and irregular with age. Length to about 20 cm.

B

C

R. nigripinnis. Off Panama, Pacific. **B** length 75 mm. **C** length 23 mm. Jerry Allen.

A

R. bistrispinus. Arraial do Cabo, Brazil. Length about 10 cm. Peter Wirtz.

Freckled Soapfish
Rypticus bistrispinus

Rypticus bistrispinus Mitchill, 1818.
Straits of Bahamas.

Western Atlantic, ranging from Florida, U.S.A. to Rio de Janeiro, Brazil. Shallow reef habitats in protected bays, resting or swimming close to substrate. Readily identified by the numerous small dusky spots over the dark-shaded upper half of the body. A brown to black stripe on snout, running through eye, gradually fading into body colour behind eye, which is most distinct in juveniles. A small species, reaching about 15 cm.

B

C

R. bistrispinus. **B** Arraial do Cabo, Brazil. Depth 4 m. L. 7 cm. Carlos Ferreira. **C** Salvador, Brazil. Depth 6 m. Bertan M. Feitoza.

White-spotted Soapfish
Rypticus maculatus

Rypticus maculatus Holbrook, 1855. Carolina, U.S.A.

U.S.A. east coast, from Rhode Island to northern Florida. A solitary species, usually seen resting on the substrate below reef-overhangs during the day. Occurs at various depths from shallow protected habitats to at least 40 m. Adults are easily identified by the white spots scattered over the sides. Length to 20 cm, but usually much smaller.

R. maculatus. Panama City, Florida. Paul Humann.

Black-spotted Soapfish
Rypticus subbifrenatus

Rypticus subbifrenatus Gill, 1861.
Virgin Is., West Indies.

Widespread West Indies, north to Florida and Bahamas and south to Rio de Janeiro, Brazil. Reef habitats with coral rubble and sponges, often seen resting on the bottom. Adults pinkish to dusky brown and easily identified by the black spots scattered over the head and body. Juveniles with black lines and spots and the median fins black with broad white margins. Length to 18 cm, but usually to 15 cm.

A

R. subbifrenatus. Adult. Dominica, Lesser Antilles. Paul Humann.

B

C

R. subbifrenatus. Dominica, Lesser Antilles. **B** juvenile. **C** subadult. Paul Humann.

D

E

R. subbifrenatus. **D** Salvador, Brazil. Depth 2 m. Bertran Feitoza. **E** Curacao, Netherlands Antilles. Peter Wirtz.

Subfamily Grammistinae, Genus *Grammistops* Schultz, 1953

Masculine. Type species: *Grammistops ocellatus* Schultz, 1953. Monotypic, widespread Indo-Pacific. Scales very small and numerous. Separate spinous and soft dorsal fins.

G. ocellatus. Cocos-Keeling Island, Indian Ocean. Length 10 cm. Jerry Allen.

False-eyed Soapfish
Grammistops ocellatus

Grammistops ocellatus Schultz, 1953.
Bikini Atoll, Pacific

Widespread in the Indo-Pacific, from the east African coast to Society Islands and Micronesia, but rare and known only from a few scattered locations. Secretive in corals, mainly protected inner reefs and lagoons. Yellowish grey to dark, blackish brown. Eye yellow and an eye-sized white-edged black ocellus on opercle. Length to about 12 cm.

Subfamily Grammistinae, Genus *Aporops* Schultz, 1943

Masculine. Type species: *Aporops bilinearis* Schultz, 1943. Monotypic genus. Similar to *Grammistops,* but lacks enlarged pores in interorbital space and has two lateral lines. Dorsal fin continuous, VII, 21.

A. bilinearis. Rowley Shoals, off north-western Australia. Length 85 mm. Jerry Allen.

Leopard Podge *Aporops bilinearis*

Aporops bilinearis Schultz, 1943. Phoenix Is.

Widespread in the Indo-Pacific, from the east African coast in the west to the Hawaiian and Marquesas Islands in the east. Occurs secretively amongst corals or rubble on inner, semi-exposed reefs and slopes below drop-offs in depths to about 30 m. Yellow-brown with many dark spots scattered over head and body. Length to 11 cm.

Subfamily Grammistinae, Genus *Suttonia* Smith, 1953

Feminine. Type species: *Suttonia suttoni* Smith, 1953. Comprises 2 species, both known only from a few specimens taken in scientific collections using chemicals. Has tiny cycloid scales; dorsal continuous, VII, 20 or 21.

S. lineata. Cocos-Keeling Island, Indian Ocean. Length 45 mm. Jerry Allen.

Freckle-face Podge *Suttonia lineata*

Suttonia lineata Gosline, 1960. Hawaiian Is.

Indo-Pacific. Scattered localities. Occurs mainly in outer reef lagoons or inner reefs, secretive in coral rubble to about 30 m depth. *Suttonia lineata* is brownish with a pale stripe from below eye to end of opercle with spots and scribbles on cheek below, and has dark shading along pectoral, dorsal, anal and caudal fin bases. *S. suttoni* Smith 1953, Mozambique, has a dark ear-spot and 3 dark longitudinal stripes on sides. Length to 9 cm.

Subfamily Grammistinae, Genus *Pseudogramma* Cuvier, 1828

Neutral. Type species: *Pseudochromis polyacantha* Bleeker, 1856. Comprises 11 small species, only two of which are in the Atlantic and others variously distributed throughout the Indo-Pacific (see list below). Very secretive fishes that are mostly collected with the use of chemicals. Although included with the soapfishes, they do not secrete toxin from the skin.

If available these fishes would do well in reef aquariums, but they are small predators and should be kept with fishes of their own size or larger. Two species are included here.

1 *Pseudogramma astigmum* Randall & Baldwin, 1997	Marshall Is.
2 *Pseudogramma australis* Randall & Baldwin, 1997	Tuamoto Archipelago
3 *Pseudogramma axilrodi* Allen & Robertson, 1995	Clipperton I., Eastern Pacific
4 *Pseudogramma erythreum* Randall & Baldwin, 1997	Philippines
5 *Pseudogramma gregoryi* (Breder, 1927)	Belize, Western Atlantic
6 *Pseudogramma guineensis* (Norman, 1935)	Gulf of Guinea, Eastern Atlantic
7 *Pseudogramma megamycterum* Randall & Baldwin, 1997	Red Sea
8 *Pseudogramma pectoralis* Randall & Baldwin, 1997	Palau
9 *Pseudogramma polyacanthum* Bleeker, 1856	Moluccas, Indonesia
10 *Pseudogramma thaumasium* (Gilbert, 1900)	Panama (Pacific)
11 *Pseudogramma xanthum* Randall & Baldwin, 1997	Tuamoto Archipelago

Bold-spot Podge *Pseudogramma polyacanthum*

Pseudochromis polyacantha Bleeker, 1856.
Moluccas, Indonesia.

Widespread Indo-Pacific, but probably a complex of similar species. Very common amongst rubble from shallow reef-flats and slopes to moderate depths, but extremely secretive and rarely seen. Yellowish grey to brown with dusky irregular blotches and a distinct black 'ear' spot. African fish have black anal-fin spines. Length to 86 mm.

A

P. polyacanthum. Flores, Indonesia. Depth 15 cm. Length 60 mm.

B

P. polyacanthum. Aliwal Shoal, South Africa. L. 40 mm.

C

P. polyacanthum. Rowley Shoals, WA, Australia. L. 6 cm. Jerry Allen.

Pacific Podge *Pseudogramma thaumasium*

Rhegma thaumasium Gilbert, 1900. Panama (Pacific).

Known from Gulf of California to Panama. Extremely secretive on rocky reefs, known only from being collected with the use of chemicals. Pale brownish, scales with dark centre forming thin longitudinal lines and similarly spaced lines on dorsal fin. A dark spot on opercle and several dusky stripes from behind and below eye. Length to 10 cm.

P. thaumasium. Panama, Pacific. Jerry Allen.

Family Serranidae, **Subfamily** Serraninae

A large subfamily comprising over 10 genera and about 80 species that are mostly confined to the Atlantic seas, 1 genus of which ranging into the Indian Ocean, 4 genera in the Eastern Pacific and 1 confined to the Indo-Pacific. Small to medium sized perch-like fishes that live on or close to reefs. Most species prey on small invertebrates or other fishes, as well as zooplankton, especially when juvenile, and some small species feed primarily on zooplankton. Most species swim close to the substrate to sneak up on prey, but many are often seen resting on the bottom. Usually those that are planktivores form small aggregations above the bottom when food is available.

Adults show no sexual dimorphism, due to the members in this subfamily being synchronous hermaphrodites. This means that an individual can function at the same time as either sex. They could fertilise their own eggs, but normally spawn in pairs and may alternate the release of eggs and sperm between spawning sessions. The hamlets (*Hypoplectrus*) have bold, often bright metallic colours, but most species in other genera have variable patterns of spots and lines or bars.

The smaller members, especially the colourful hamlets, are popular with aquarists, but being predators and often aggressive (especially to their own kind), selection should be carefully considered for the community tank. Serranids are generally easily kept and accept most foods offered. The larger members, such as those in the genus *Paralabrax* or the European members of *Serranus*, are caught on line or netted and are considered to be good to eat.

Genera Contents

SUBFAMILY SERRANINAE, **GENUS** *Serranus* Steindachner, 1874

Masculine. Type-species: *Perca cabrilla* Linnaeus, 1758. Comprises 30 described and several undescribed species, most of which occur in the Atlantic seas and the rest are known from the Eastern Pacific and south-western Indian Ocean. A few species in the West Indies are known only from type-material, but may be synonymous with other species. There are 9 or 10 species recognised in the eastern Atlantic and Mediterranean seas, 2 in the southern and western Indian Ocean and 7 in the Eastern Pacific.

Predatory fishes feeding on a variety of swimming or crawling prey from zooplankton to shrimps, worms and fishes. Most species in the Americas are small, ranging from 6 to about 20 cm maximum, but in Europe most species reach 35 to 40 cm and are primarily used as food fish.

1 *Serranus accraensis* (Norman, 1931)	W. Africa	n.i.
2 *Serranus aequidens* Geoffroy Saint-Hilaire, 1817	Cal., Mex.	n.i.
3 *Serranus africanus* (Cadenat, 1960)	Nigeria	n.i.
4 *Serranus annularis* (Günther, 1880)	Brazil	166
5 *Serranus atricauda* Günther, 1874	Morocco	158
6 *Serranus atrobranchus* (Cuvier, 1829)	Brazil	n.i.
7 *Serranus baldwini* (Evermann & Marsh, 1899)	Puerto Rico	165
8 *Serranus cabrilla* (Linnaeus, 1758)	Europe	156
9 *Serranus capurri* Perugia, 1896	West Indies	n.i.
10 *Serranus chionaraia* Robins & Starck, 1961	Florida	n.i.
11 *Serranus flavescens* Valenciennes, 1830	West Indies	n.i.
12 *Serranus flaviventris* (Cuvier, 1829)	Brazil	163
13 *Serranus hepatus* (Linnaeus, 1758)	Mediterranean	159
14 *Serranus huascarii* Steindachner, 1900	Peru	n.i.
15 *Serranus knysnaensis* Gilchrist, 1904	South Africa	157
16 *Serranus luciopercanus* Poey, 1852	Cuba	n.i.
17 *Serranus maytagi* Robins & Starck, 1961	Caribbean	n.i.
18 *Serranus notospilus* Longley, 1935	Florida	166
19 *Serranus novemcinctus* Kner, 1864	St Paul (Indian Ocean)	n.i.
20 *Serranus phoebe* Poey, 1851	Cuba	169
21 *Serranus psittacinus* Valenciennes, 1830	Galápagos	164
22 *Serranus sanctaehelenae* Boulenger, 1895	Saint Helena	n.i.
23 *Serranus scriba* (Linnaeus, 1758)	Mediterranean	160
24 *Serranus socorroensis* Allen & Robertson, 1992	E. Pacific	166
25 *Serranus stilbostigma* (Jordan & Bollman, 1890)	Galápagos	n.i.
26 *Serranus subligarius* (Cope, 1870)	Florida	163
27 *Serranus tabacarius* (Cuvier, 1829)	West Indies	168
28 *Serranus tico* Meisler & Lavenberg, 1995)	E. Pacific	166
29 *Serranus tigrinus* (Bloch, 1790)	West Indies	167
30 *Serranus tortugarum* Longley, 1935	Florida	168
31 *Serranus* sp	São Tomé	162

n.i. = not included

Serranus cabrilla. Type-species for genus. Gran Canaria, Canary Islands. Peter Wirtz.

Barred Comber *Serranus cabrilla*

Perca cabrilla Linnaeus, 1758. No locality.

Widespread Mediterranean and Atlantic, north to English Channel, Azores and subtropical west African coast. Lives in rocky habitats, but may visit seagrasses nearby. Seen at various depths and reported to 600 m. Juveniles feature two black horizontal stripes that break up into blotches with growth and develop a barred pattern. Adults have 7–9 reddish brown bars on the body. Deep water individuals are more red or yellow. Length to 30 cm, rarely 40 cm.

A

S. cabrilla. Western Mediterranean. D. 13 m. L. 20 cm. Patrick Louisy.

B

C

S. cabrilla. Corsica, Mediterranean. **B** depth 6 m. Length 5 cm. **C** depth 15 m. Length 35 mm. Patrick Louisy.

D

E

S. cabrilla. Mediterranean. **D** Banyuls. Depth 15 m. Length 15 cm. **E** Cerbére. Depth 6 m. Length 16 cm. Patrick Louisy.

F

G

S. cabrilla. Mediterranean. Night colours. **F**. Banyuls. Depth 8 m. L. 13 cm. **G** Cerbére. Depth 4 m. L. 16 cm. Patrick Louisy.

S. cabrilla. **H** aquarium, aggressive display. Mediterranean. Patrick Louisy. **D** adult from deep water. Azores. Peter Wirtz.

S. cabrilla. Western Mediterranean. Depth 22 m. Length 25 cm. Patrick Louisy.

Knysna Comber *Serranus knysnaensis*

Serranus knysnaensis Gilchrist, 1904. Knysna, Cape Provence, South Africa.

South Africa, ranging to KwaZulu-Natal on the east coast. Shallow rocky reefs in protected bays from about 10 m down, deeper offshore from about 25 m down. Juveniles and subadults pale, brown above and white below, with longitudinal dark stripes and adults darker with additional, but indistinct barred pattern. Caudal fin with dusky to red tips. Length to 20 cm.

S. knysnaensis. Krysna, South Africa. Adult. Guido Zsilavect.

S. knysnaensis. KwaZulu-Natal, South Africa. Depth 25 m. **B** juvenile. **C** & **D** subadults. Dennis King.

Black-tipped Comber *Serranus atricauda*

Serranus atricauda Günther, 1874. Morocco.

Eastern Atlantic, ranging from Gulf of Biscay to Mauretania, into western part of the Mediterranean and occurs in the Azores and the Canary Islands. Rocky reefs with coralline algae coverage at a depth range of about 10–50 m but reported to 90 m. Juveniles with two black parallel stripes and grey to brownish backs. Stripes remain solid until subadult size and then a barred pattern develops that in turn changes to a series of dark blotches. Caudal fin tips black in large adults. Length to 30 cm, rarely 35 cm.

Remarks: juveniles and sub-adults similar to the closely related *Serranus knysnaensis* from south Africa, a species that was confused with *S. cabrilla*.

A

S. atricauda. Large adult. Madeira. Peter Wirtz.

B

C

D

S. atricauda. **B** juvenile. Madeira. **C** & **D** adults. Azores. Peter Wirtz.

E

S. atricauda. Adults. Azores. Peter Wirtz.

Brown Comber *Serranus hepatus*

Labrus hepatus Linnaeus, 1758.
Mediterranean Sea.

Widespread in the Mediterranean Sea and eastern Atlantic ranging from Portugal to Angola. Also found in the Azores and Canary Islands. Various reef habitats from shallow protected waters to depths of several hundred metres. This species is easily identified by the dark bands on the body and the black spot in the dorsal fin. Length 20 cm, rarely 25 cm.

A

S. hepatus. Adult in aquarium. Length 16 cm. Frank Schneidewind.

B

C

S. hepatus. Golfe du Lion, Mediterranean. **B** length 15 cm. **C** depth 12 m. Length 6 cm. Patrick Louisy.

D

E

S. hepatus. Golfe du Lion, Mediterranean. Depth 15 m. Length 35 mm. Patrick Louisy.

F

G

S. hepatus. Golfe du Lion, Mediterranean. **F** depth 14 m. Length 5 cm. **G** hiding. Depth 15 m. Length 6 cm. Patrick Louisy.

Painted Comber *Serranus scriba*

Perca scriba Linnaeus, 1758. No locality.

Widespread Mediterranean and eastern Atlantic, north to English Channel, and south to Mauritania, African coast, and Canary Islands. Lives in mixed rocky and algae habitats, but ventures into seagrass beds. Shallow protected areas to about 30 m depth, reported to 150 m. Easily identified by its distinctive barred pattern, and especially during courtship when abdomen is blue. Length to 30 cm, rarely 36 cm.

Remarks: the colour pattern in the Canary Islands is considerably different and may warrant investigation.

A

S. scriba. Provence. Mediterranean Sea. D. 10 m. L. 22 cm. Patrick Louisy.

B

S. scriba. Italy. Mediterranean Sea. Kai Velling.

C

D

S. scriba. Golfe du Lion, Mediterranean Sea. **C** courting. Depth 18 m. Length 24–25 cm. **D** depth 6 m. L. 20 cm. Patrick Louisy.

E

S. scriba. Adult colour variation. Tenerife, Canary Islands. Peter Wirtz.

F

S. scriba. Golfe du Lion, Mediterranean Sea. Courting. Depth 18 m. Length 24–25 cm. Patrick Louisy.

G

São Tomé Comber *Serranus* sp

Undetermined, apparently undescribed.

Known only from islands in the Gulf of Guinea, off the tropical coast of west Africa. A solitary species found on rocky reefs at about 10 m depth, often resting on substrate and may look more like a scorpaenid than a serranid. Variable in colour, mostly red with pale bars or horizontal stripes, and distinct white markings under the head and chest area (**B**). Length to about 15 cm.

A

S. sp. São Tomé, Gulf of Guinea. Length ~15 cm. Peter Wirtz.

B

C

S. sp. São Tomé, Gulf of Guinea. Peter Wirtz.

D

S. sp. São Tomé, Gulf of Guinea. Peter Wirtz.

Twinspot Bass *Serranus flaviventris*

Dules flaviventris Cuvier, 1829. Brazil.

Western Atlantic, Caribbean to Uruguay. Reef habitats. Shallow protected bays and deep offshore to at least 400 m. A solitary species, usually seen in or at entrance of crevices. A distinctive species due to its moderately deep body and black spots in the dorsal and caudal fins, latter as twin spots near its base that gave rise to its English name. In Brazil it is known as Mariquita or Mariquinha. A smallish species, length to about 15 cm.

S. flaviventris. Mochima National Park, Venezuela. Paul Humann.

S. flaviventris. Abrolhos I., Brazil. **B** depth 5 m. Length 6 cm. Carlos Ferreira. **C** depth 10 m. Length 11 cm. Jerry Allen.

Belted Sand-bass *Serranus subligarius*

Centropristis subligarius Cope, 1870. Florida.

Mainly known from Florida to north Carolina, and ranging into the northern part of Gulf of Mexico. Occurs on open sandy and rubble substrates with rocky outcrops to about 30 m depth. Identified by its generally dusky appearance, a white or bluish white blotch covering most of the abdomen, and the black bands on posterior half of the body, one of which extending into the dorsal fin, where forming a large tear-shaped black blotch. A small species, length to 10 cm.

S. subligarius. Aquarium. Scott Michael.

S. subligarius. **B** Jacksonville, Florida. Paul Humann. **C** North Carolina, U.S.A. Peter Wirtz.

Galápagos Bass *Serranus psittacinus*

Serranus psittacinus Valenciennes, 1855. Galápagos.

Eastern Pacific, Gulf of California to Peru and offshore islands, including Galápagos. Mainly found on patch reef on open sandy or rubble substrates from the shallows to depths of at least 60 m. Juvenile pale, grey on top, white below, with a black mid-lateral stripe. Adults with barred pattern over the back, the grey to brown-red stripes in pairs, and a series of spots along lower sides. Length to 12 cm.

Remarks: often referred to as *Serranus fasciatus*, an older but unavailable name.

A

S. psittacinus. I. Taboguilla, Panama. Jerry Allen.

B

C

D

S. psittacinus. I. Uva, Panama. Jerry Allen.

E

S. psittacinus. Juv. Galápagos. Frank Schneidewind.

F

S. psittacinus. Pinzon, Galápagos. Length 12 cm. Jerry Allen.

Lantern Bass *Serranus baldwini*

Prionodus baldwini Evermann & Marsh, 1899.
Puerto Rico.

Tropical Western Atlantic, ranging north to south Florida and Bahamas, and south to São Paulo, eastern Brazil and offshore islands. Various reef habitats to 70 m depth. Often resting on the bottom and easily approached, often curious towards divers. Identified by the pattern of blotches and lines, especially the black blotches along lower sides and caudal-fin base. Length to 6 cm in the West Indies, but can reach 10 cm in their southern range of Brazil.

A

S. baldwini. Bonaire, Netherlands Antilles. D. 12 m. L. 6 cm. Roger Steene.

B

C

S. baldwini. Arraial do Cabo, Brazil. Peter Wirtz.

D

S. baldwini. Cozumel. D. 12 m. Length 55 mm. Roger Steene.

E

S. baldwini. Abrolhos I., Brazil. Depth 10 m. Length 8 cm. Jerry Allen.

S. tico. Isla del Coco, Eastern Pacific. Length 6 cm. Ross Robertson.

Red-head Sand-bass *Serranus tico*

Serranus tico Meisler & Lavenberg, 1995. Isla del Coco, Costa Rica, Eastern Pacific.

Known only from Cocos Islands. Occurs on sand and rubble around reefs. Mostly brown-red, body with white bars over abdomen and alternating with red over posterior half of body, a broad white section on caudal peduncle as if one red bar is missing. A small species, length to 7 cm.

S. socorroensis. Isla Socorro, Eastern Pacific. Length 6 cm. Jerry Allen.

Revillagigedos Bass *Serranus socorroensis*

Serranus socorroensis Allen & Robertson, 1992. Isla Socorro, Islas Revillagigedos, Eastern Pacific.

Known only from the Revillagigedos Islands in the Eastern Pacific. Occurs on shallow sand flats with scattered rocks and corals. The type specimens were collected at a depth of 8 m. A small species, mostly red with a series of pale blotches along the back and lower sides. Length to 65 mm.

S. annularis. West Palm Beach, Florida. Length about 5 cm. Paul Humann.

Orange-back Bass *Serranus annularis*

Centropristis annularis Günther, 1864. Cuba.

Western Atlantic, Florida, Bermudas and Bahamas to the Caribbean Sea. Occurs around rocky reefs, usually near crevices or below overhangs and at depths of about 30 m or more, occasionally shallower. Brown to orange, series of white blotches forming a band mid-laterally and two distinct orange blotches outlined by dark borders behind eye. Small species, length to about 6 cm, but usually smaller.

S. notospilus. Isla Cubagua, Venezuela. Depth 7 m. Paul Humann.

Striped Sand-bass *Serranus notospilus*

Serranus notospilus Longley, 1935. Florida.

Florida and Gulf of Mexico to Surinam. Found on sand and rubble adjacent to reefs in a depth range of about 5–30 m. Pale with three longitudinal lines along upper half of the body, lower mid-laterally and running at eye level, ending in distinct spot on caudal-fin base. Length to 12 cm.

Harlequin Bass *Serranus tigrinus*

Holocentrus trigrinus Bloch, 1790. Indis.

Western Atlantic, south Florida, Bermudas and Bahamas to widespread in the Caribbean Sea as far as Lesser Antilles. Various habitats of mixed reef and seagrasses from shallow protected areas to about 40 m depth. Usually seen swimming slowly near the bottom looking out for prey such as small crustaceans. Readily identified by its distinct pattern of spots, blotches or bars. Length to about 10 cm

A

S. tigrinus. Aquarium. Length 8 cm.

B

S. tigrinus. Lesser Antilles. Juvenile, 55 mm. Patrick Louisy.

C

S. tigrinus. Curacao. Peter Wirtz.

D

S. tigrinus. Pair prior to spawning. Lesser Antilles. Depth 7 m. Length 8 cm. Patrick Louisy.

E

S. tigrinus. St. Vincent. Depth 10 m. Length 10 cm. Roger Steene.

Tobaccofish *Serranus tabacarius*

Centropristes tabacarius Cuvier, 1829.
Martinique I., West Indies.

Widespread in the Caribbean Sea, south Florida and Bahamas, ranging to northern Brazil. Mixed reef and sand habitat from shallow bays to at least 70 m deep. Uniquely coloured with ochre sides, dark and pale blotched pattern over the back, and dark streaks along outer lobes of caudal fin. Moves about slowly and often curious towards divers. One of the larger species in this genus in the western Atlantic, length to about 18 cm.

A

S. tabacarius. Aquarium. Length 15 cm.

B

S. tabacarius. Bonaire. Length 12 cm. Roger Steene.

C

S. tabacarius. Lesser Antilles. Length 7 cm. Patrick Louisy.

Chalk Bass *Serranus tortugarum*

Serranus tortugarum Longley, 1935. Florida.

Western Atlantic from south Florida and Bahamas to the central Caribbean Sea region. Swims in small groups on open substrates with coral outcrops, usually in depths over 10 m. Reported to 100 m. Easily identified by the blue sides and dark blotches over the back. Length to 9 cm, usually to 7 cm.

Remarks: popular and easily kept aquarium species that is usually recommended for beginners.

A

S. tortugarum. Aquarium. Length 7 cm. Scott Michael.

B

S. tortugarum. Aquarium. Length 6 cm.

C

S. tortugarum. Florida. Depth 8 m Length 6 cm. Roger Steene.

Tattler Bass *Serranus phoebe*

Serranus phoebe Poey, 1851. Cuba.

Western Atlantic, known from Florida and south Carolina, Bermudas and Gulf of Mexico, ranging south to Santa Catarina, Brazil. Occurs on sand and low reef habitats, and usually in depths over 30 m. Reported to a depth of 400 m. A distinctive species, generally a sandy colour with dark bands. Juveniles have one band angling back from below eye, a second from anterior dorsal spines to abdomen that fades with age. Adults with longitudinal stripes and black blotch anteriorly on the dorsal fin that was part of the juvenile band. Length to 20 cm.

A *S. phoebe.* Juvenile. West Palm Beach, Florida. Paul Humann.

B *S. phoebe.* Arraial do Cabo, Brazil. Peter Wirtz.

C *S. phoebe.* Arraial do Cabo, Brazil. Length about 18 cm. Peter Wirtz.

Subfamily Serraninae, Genus *Chelidoperca* Boulenger, 1895

Feminine. Type-species: *Centropristis hirundinaceus* Weber, 1913. A small Indo-Pacific genus of deep water basses, comprising 6 species known as sand perches. Fishes with elongate and moderately compressed body; large, dorsally placed eyes; mouth large with villiform teeth in jaws; single dorsal fin with shallow notch. Bottom dwelling fishes, usually resting on substrate and perched on ventral fins. Sand perches are superficially much like grubfishes (Mugiloididae, *Parapercis*). Caught by trawl, usually in depths of 100 m or more, but occasionally entering shallower depths. Easily kept in the aquarium (cool water) if brought up successfully from the depth.

1 *Chelidoperca hirundinacea* (Valenciennes, 1831) Japan
2 *Chelidoperca investigaris* (Alcock, 1890) India
3 *Chelidoperca lecromi* Fourmanoir, 1982 Coral Sea
4 *Chelidoperca margaritifera* Weber, 1913 Japan to New Guinea
5 *Chelidoperca occipitalis* Kotthaus, 1973. Arabian Sea
6 *Chelidoperca pleurospilus* (Günther, 1880) Southern Indonesia

C. hirundinacea. Type-species for genus. Tokai Aquarium, Japan. Length 16 cm.

Black-spot Sand-perch
Chelidoperca hirundinacea

Centropristis hirundinaceus Valenciennes, 1831. Japan.

Southern Japan to South China Sea. Lives on sandy substrates, coastal, somewhat silty habitat in a depth range of 80–200 m. Pink, often with many yellow stripes along vertical scale-rows and an eye-sized dark spot centrally just below lateral line. Length to about 22 cm.

Subfamily Serraninae, Genus *Serraniculus* Ginsburg, 1952

Masculine. Type-species: *Serraniculus pumilio* Longley, 1935. A monotypic western Atlantic genus.

S. pumilio. Panama City, Florida. Length about 6 cm. Paul Humann.

Pygmy Sea-bass
Serraniculus pumilio

Serraniculus pumilio Longley, 1935.

Known from north Carolina to Venezuela. Mainly found in continental waters, in habitats of mixed sand and rubble, in the shallows with sparse seagrasses. Reported to a depth of 70 m. Pale, sandy coloured, with dark blotches, reddish markings and several black spots over the back at dorsal-fin base. A broad pale band on the caudal peduncle. Easily mistaken for a small scorpionfish. Length to 6 cm.

SUBFAMILY SERRANINAE, GENUS *Dules* Cuvier, 1829

Masculine. Type-species: *Dules auriga* Cuvier, 1829. A monotypic genus from Brazil. *Centropristis brasiliensis* Brisout de Barneville, 1847, is also this species.

Longfin Sea-bass *Dules auriga*

Dules auriga Cuvier, 1829. Brazil.

Western Atlantic, Brazil and south to northern Argentina. Occurs on sand and rubble substrates adjacent to reefs. Depth range 15 to 140 m, most common over 30 m. Readily identified by its long third dorsal-fin spine and colouration. Pale, sandy coloured with black markings, a V-shaped band from dorsal to anal fin, ventral fins black and a black area adjacent to the fins on the abdomen. Length to about 23 cm.

A

B

D. auriga. Brazil. **A** length 16 cm. **B** depth 35 m. Length 16 cm. Alfredo Carvalho.

SUBFAMILY SERRANINAE, GENUS *Schultzea* Woods, 1958

Feminine. Type-species: *Schultzea campechanus* Woods, 1958 (=*Serranus beta* Hildebrand, 1940). A monotypic genus, represented by a small, wrasse-like deep water planktivore in the western Atlantic.

Schooling Pixie Bass *Schultzea beta*

Serranus beta Hildebrand, 1940. Florida.

Known from off Florida and West Indies. Adults mainly known from deep water to about 100 m depth. Schools above rocky substrate to feed on plankton. Juveniles pinkish with several dusky lines. Adult with blotched pattern. Length to 10 cm.

A

B

S. beta. Curacao, Netherlands Antilles. Peter Wirtz.

Subfamily Serraninae, **Genus** *Centropristis* Cuvier, 1829

Feminine. Type-species: *Centropristis nigricans* Cuvier, 1829 (= *Coryphaena nigrescens* Bloch & Schneider, 1801, = *Labrus striatus* Linnaeus, 1758). Comprises 4 Western Atlantic species. Some grow to a large size, reaching 60 cm and a weight of about 5 kg and are commercial species.

1 *Centropristis fuscula* Poey, 1861	Cuba
2 *Centropristis ocyurus* (Jordan & Evermann, 1887)	Florida
3 *Centropristis philadelphica* (Linnaeus, 1758)	America
4 *Centropristis striata* (Linnaeus, 1758)	America

Black Sea-bass *Centropristis striata*

Labrus striatus Linnaeus, 1758. America.

Western Atlantic, mainly known from continental waters and ranging from Massachusetts to Florida and Gulf of Mexico. Rocky reefs, often common under jetties and along sea-walls, ranging to a depth of about 40 m. Juveniles very dark over the back and pale below, a black mid-lateral stripe from behind eye and a black spot centrally on dorsal fin base. Dorsal and outer margins of caudal fin white. Adults with similar pattern but less distinct and additional barring. A large species, usually to 45 cm, but can reach 60 cm. They are reputed to be long lived and can reach 20 years of age, but fishing pressure may have now made such individuals very rare.

Remarks: *Centropristis rufus* Cuvier, 1829, is regarded as a synonym of this species.

A

C. striata. Juvenile. Jacksonville, Florida. Length about 9 cm. Paul Humann.

B

C. striata. Adult. Jacksonville, Florida. Length about 35 cm. Paul Humann.

Bank Sea-bass *Centropristis ocyurus*

Holocentrus ocyurus Jordan & Evermann, 1887. Florida, U.S.A.

Western Atlantic, from North Carolina south to Florida and into Gulf of Mexico. Occurs on sand and rubble adjacent to reefs in a depth range of about 20–80 m. Juveniles pale with a series of elongate black blotches forming a mid-lateral stripe. The stripe breaks up into spots with growth and a barred pattern develops. Head often yellowish. Caudal fin in large adults with extended rays on tips and in centre. Length to 30 cm, but rarely over 25 cm.

C. ocyurus. West Palm Beach, Florida. Length about 5 cm. Paul Humann.

C. ocyurus. West Palm Beach, Florida. Length about 20 cm. Paul Humann.

C. ocyurus. Panama City, Florida. Length about 28 cm. Paul Humann.

SUBFAMILY SERRANINAE, GENUS *Paralabrax* Gerard, 1856

Masculine. Type-species: *Labrax nebulifer* Gerard, 1854. Confined to the Americas, comprising 9 species, divided between the western Atlantic and the eastern Pacific. Reef dwellers, but usually on margins with sand and rubble zones. Medium sized fishes, ranging from about 30 to 60 cm in length.

1 *Paralabrax albomaculatus* (Jenyns, 1840)	Galápagos Islands
2 *Paralabrax auroguttatus* Walford, 1936	West Mexico
3 *Paralabrax callaensis* Starks, 1906	Peru
4 *Paralabrax clathratus* (Girard, 1854)	Venezuela
5 *Paralabrax dewegeri* (Metzelaar, 1919)	Venezuela
6 *Paralabrax humeralis* (Valenciennes, 1828)	Chile
7 *Paralabrax loro* Walford, 1936	West Mexico
8 *Paralabrax maculatofasciatus* (Steindachner, 1900)	West Mexico
9 *Paralabrax nebulifer* (Girard, 1854)	California

P. albomaculatus. Large adult. Galápagos Islands. Length ~30 cm. Jerry Allen.

White-blotched Sea-bass
Paralabrax albomaculatus

Serranus albomaculatus Jenyns, 1840. Galápagos Islands.

Known only from the Galápagos Islands. Usually found on rocky, low-profile reefs and nearby sand patches at a depth range of about 10–30 m. Identified by the series of eye-sized white blotches mid-laterally and above. Length to about 38 cm.

Remarks: may be synonymous with the very similar Peruvian Sea-bass *Paralabrax humeralis* (Valenciennes, 1828).

P. maculatofasciatus. Sea of Cortez. Roger Steene.

Spotted Sea-bass
Paralabrax maculatofasciatus

Serranus maculatofasciatus Steindachner, 1868. W. Mexico.

Eastern Pacific, ranging from central California to Gulf of California. Usually found on rocky, low-profile algae-covered reefs and nearby sand patches in a depth range of about 1–60 m. Identified by the combination of dark spots and bars. A large species, length to about 56 cm.

P. dewegeri. West Palm Beach, Florida. Length about 20 cm. Paul Humann.

Deweger's Sea-bass
Paralabrax dewegeri

Serranus dewegeri Metzelaar, 1919. Venezuela.

Caribbean Sea to Venezuela and northern Brazil. Occurs mainly on sand or rubble patches amongst or near rocky reefs. Third and fourth dorsal-fin spines longest. Body dark with pale blotches on upper sides and barring on lower sides. A large species, reaching at least 40 cm in length and well over 1 kg in weight. Largest known is 1.3 kg.

Subfamily Serraninae, **Genus** *Diplectrum* Holbrook, 1855

Neutral. Type-species: *Serranus fasicularis* Valenciennes, 1828 (= *Perca formosa* Linnaeus, 1766). Comprises 12 species divided between the western Atlantic and eastern Pacific. Primarily occur in relatively deep water or over open sand flats and are rarely seen or photographed. One species (type-species) is included here.

1 *Diplectrum bivittatum* (Valenciennes, 1828)	Martinique I., West Indies
2 *Diplectrum conceptione* (Valenciennes, 1828)	Chile
3 *Diplectrum eumelum* Rosenblatt & Johnson, 1974	Jalisco, Mexico
4 *Diplectrum euryplectrum* Jordan & Bollman, 1890	Colombia, Pacific
5 *Diplectrum formosum* (Linnaeus, 1766)	Carolinas, U.S.A.
6 *Diplectrum labarum* Rosenblatt & Johnson, 1974	Jalisco, Mexico
7 *Diplectrum macropoma* (Günther, 1846)	Panama, Pacific
8 *Diplectrum maximum* Hildebrand, 1946	Peru
9 *Diplectrum pacificum* Meek & Hildebrand, 1925	Panama, Pacific
10 *Diplectrum radialis* (Quoy & Gaimard, 1824)	Brazil
11 *Diplectrum rostrum* Bortone, 1974	Ecuador, Pacific
12 *Diplectrum sciurus* Gilbert, 1892	Gulf of California

Sand Sea-bass
Diplectrum formosum

Perca formosa Linnaeus, 1766. Carolinas.

Western Atlantic, from Virginia to Florida and Gulf of Mexico, ranging south to São Paulo, Brazil and also in Bermudas. Occurs on sand and seagrass habitats adjacent to reefs from shallow depths to at least 70 m. Pale, yellowish with blue longitudinal lines and a series of dark blotches that may form vertical bands. Length to 30 cm, usually to 25 cm.

Remarks: two forms are recognised, southern form has fewer gill rakers and is smaller than northern one.

A *D. formosum*. West Palm Beach, Florida. Depth 20 m. Paul Humann.

B *D. formosum*. Florida. Depth 9 m. Paul Humann.

C *D. formosum*. Florida. Depth 5 m. Roger Steene.

Subfamily Serraninae, Genus *Hypoplectrus* Gill. 1861

Masculine. Type-species: *Plectropoma puella* Cuvier, 1829. A tropical western Atlantic genus with 13 species that are virtually identical in morphology and identified by colour. They are commonly called 'hamlets' and the various species appear to be closely related to each other and may have only evolved recently from a deep-bodied *Serranus*-like sea-bass. Some species are highly variable in colour and several species appear to mimic damselfishes of the genus *Stegastes*. This probably enables them to approach unsuspecting prey such as small fishes or shrimps that do not expect problems from algae-feeding damsels. Some species are thought to hybridise. As synchronous hermaphrodites, adults can function as either sex at the same time. They could fertilise their own eggs, but normally spawn in pairs.

In the aquarium

Many species are attractive, often with bright metallic colours, and with a maximum size of about 14 cm, they are ideal candidates for the aquarium trade. Unfortunately they are often aggressive towards their own kind or similar species and may eat smaller fishes. On their own, they are usually easily kept and will readily accept various chopped seafood. Preferably diet is varied (true for all fishes) to keep them healthy the long-term. They should be given lots of space to swim in as well as plenty of places to hide in or swim through. Juvenile specimens will normally easily adjust to captivity, but large specimens, collected from the wild, could be difficult and are best avoided when starting off. If several individuals are desired, they should be of similar size and introduced to their new home simultaneously. When introducing a hamlet to a tank containing an established one, the may be newcomer will almost certainly be confronted by an angry individual and this may be disastrous. In such a case it maybe necessary to have a glass or perspex partition so they can get used to each other first.

Reference: Michael, Scott W., 1998. Reef fishes Volume 1. Microcosm. Vermont.

Hypoplectrus puella. Type-species for genus. Bonaire. Depth 10 m. Roger Steene.

Hypoplectrus puella. Bonaire. Depth 12 m. Roger Steene. Courtship and spawning. Above: an acting male fully coloured and fins raised. Right and below: sequence of events leading up to spawning, twisting around each other, the acting male nudging the acting female whilst producing a short series of bongo-like sounds to stimulate its partner to release eggs.

Hypoplectrus unicolor. Bonaire. Depth 10 m. Roger Steene. Courtship and spawning. Above left: the acting male and female swimming side by side. Above right and below: courtship behaviour leading up to spawning. Acting male nudges the swollen abdomen of the acting female and to encourage egg-release it produces a drum-like low frequency sound, which is barely detectable by the human ear.

Sounds are produced by many other fishes, such as large angels (Pomacanthidae) to startle a possible enemy, and a small, 8 cm species from southern Australia, a sawbelly (Trachichthyidae) was named *Aulotrachichthys pulsator* for the loud pulses it produced in territorial disputes. In this species the sound frequency is about 1000 Hz, whilst in the Hamlet it is about 500 Hz. The noise, made by a muscle on the swimbladder, is produced in short, drum-like repeated bursts a sound. In the case of the little sawbelly, it sounds much like hitting a cup repeatedly with a metal spoon.

Masked Hamlet *Hypoplectrus providencianus*

Hypoplectrus providencianus Acero & Garzón-Ferreira, 1994. I. Providencia, Caribbean Sea.

Known only from the eastern half of the Caribbean Sea from the Providence Island to Belize and Cayman Islands. Occurs on reefs with hard corals to depths of about 15 m. Identified by its pale, bluish grey colour and distinctive black marking over the eye, extending down, usually outlined by a thin blue margin, and black markings on caudal fin base and outer lobes. Length to 13 cm.

A

B

H. providencianus. Cayman I. D. 15 m. L. 75 mm. Roger Steene.

Yellow-tail Hamlet *Hypoplectrus chlorurus*

Plectropoma chlororum Cuvier, 1828. Caribbean Sea.

Mainly found in the Lesser and Netherlands Antilles, but ranges along the Atlantic margin to the Bahamas. Reef habitats with hard corals from the shallows to at least 30 m depth. Dark brown to almost black with a pale to vivid yellow tail, but colour varies geographically and occasionally with blue scribbles or lines on head. Length to 13 cm.

A

B

H. chlorurus. St. Vincent. D. 15 m. Roger Steene.

H. sp. Aquarium. A possible hybrid, parent species thought to be *H. gemma* and *H. puella*. Scott Michael.

Barred Hamlet *Hypoplectrus puella*

Plectropoma puella Cuvier, 1828.
Martinique I., West Indies.

Most widespread Hamlet in the Caribbean, ranging into the Gulf of Mexico and into Atlantic to Bermuda. Various reef habitats, often in more turbid waters than other hamlets, ranging to a depth of about 30 m. Highly variable in colour and can change quickly with mood. Normally with barred pattern on body and head, and fine scribble lines on most of the dorsal fin. Length to about 13 cm.

Remarks: apparently hybridises with several other hamlet species.

A *H. puella*. St. Vincent. Depth 12 m. Roger Steene.

B *H. puella*. Lesser Antilles. Depth 6 m. Length 5 cm. Patrick Louisy.

C *H. puella*. Juveniles. Bonaire. Depth 12 m. Roger Steene.

D *H. puella*. Guadeloupe, Lesser Antilles. Depth 15 m. Length 7 cm. Patrick Louisy.

Blue-lined Yellow-belly Hamlet *Hypoplectrus maculiferus*

Hypoplectrus maculiferus Poey, 1871.
Havana, Cuba.

Eastern Caribbean, Greater Antilles and Virgin Islands. Reef habitats to about 15 m depth. Mostly yellow, dark bluish above and blue spots and scribbles on the head. Latter are features similar to *Hypoplectrus puella*, but *H. maculiferus* lacks banding on the body. It may occasionally hybridise with the former and other species. Length to 13 cm.

Remarks: at some stage (Domeier, 1994) it was suggested to be a possible synonym of *Hypoplectrus aberrans* and treated as such by some recent authors.

H. maculiferus. Cayman. Depth 8 m. Paul Humann.

Indigo Hamlet *Hypoplectrus indigo*

Plectropoma indigo Poey, 1851.
Havana, Cuba.

Mainly in the Greater Antilles, but occasionally found off Florida. Reef habitats, usually a little deeper than other hamlets, at depths over 10 m, ranging to about 40 m. Mainly pale or dark blue with narrow pale bands near and on head and posterior part of body and caudal-fin base. Similar pattern to *Hypoplectrus puella*, but distinguished from it by the strong, overall blue colour. Length to about 14 cm.

A

H. indigo. Aquarium. Frank Schneidewind.

B

H. indigo. Aquarium. Scott Michael.

Blue Hamlet *Hypoplectrus gemma*

Hypoplectrus gemma
Goode & Bean, 1882. Florida.

Known mainly from Florida, where common, and Belize. Occurs on shallow reefs with rich sessile invertebrate growth to about 12 m depth. Readily identified by the all-over metallic-blue colour and black streaks along outer margins of caudal fin. Caudal fin sometimes pale yellowish. Length to 13 cm.

A

H. gemma. Florida Keys. Depth 10 m. Roger Steene.

B

H. gemma. Aquarium. Length 10 cm.

C

H. gemma. Aquarium. Frank Schneidewind.

Butter Hamlet *Hypoplectrus unicolor*

Perca unicolor Walbaum, 1792. No locality.

Widespread Caribbean and most common hamlet in the area. Various reef habitats, shallow to depths of about 25 m, and found from rich invertebrate environments to human made structures. Colour pale yellow with a conspicuous black peduncular spot and a black spot on the snout that appears intermittently with mood. Length to 13 cm.

A

H. unicolor. Bonaire. Depth 10 m. Length 12 cm. Roger Steene.

B

H. unicolor. Aquarium. Length 12 cm.

C

H. unicolor. Aquarium. Length 10 cm. Frank Schneidewind.

Black Hamlet *Hypoplectrus nigricans*

Plectropoma nigricans Poey, 1852.
Havana, Cuba.

Widespread in the Caribbean, most common in the Virgin Islands and Puerto Rico regions. A very shallow water species, mainly at depths of less than 10 m, but occasionally deeper depending on conditions and locality. Dark brown to bluish or almost black. Usually a small blue 'ear' spot just below origin of lateral line. Grows a bit larger than most hamlets, reaching 15 cm.

A

H. nigricans. Lesser Antilles. Depth 15 m. Length 10 cm. Patrick Louisy.

B

H. nigricans. Cayman I. Depth 16 m. Length 7 cm. Roger Steene.

C

H. nigricans. Florida Keys. Depth 12 m. Roger Steene.

Tan Hamlet *Hypoplectrus* sp

Undetermined species.

Known from Florida and Panama. This species is well-known, but still apparently undescribed taxon. Occurs on shallow reefs to about 15 m depth. A plain species, mostly brown with a distinctive yellow eye. Scales on anterior part of body with tiny yellow spots. Lips often grey or bluish, especially on lower jaw, and in some literature has been referred to as Blue-lip Hamlet. Length to 13 cm.

H. sp. Florida Keys. Depth 8 m. Roger Steene.

Shy Hamlet *Hypoplectrus guttavarius*

Plectropoma guttavarium Poey, 1852. Havana, Cuba.

The Greater and Lesser Antilles, ranging to Florida, Bahamas and Honduras. Reef habitats to about 15 m depth in the Caribbean, but usually deeper in the Atlantic part of its range. A primarily yellow fish with dark sides from below dorsal-fin base, variably reaching ventrally from just a saddle-like streak to almost the whole side. Snout with black spot and blue markings. Length to 13 cm.

A

H. guttavarius. Florida Keys. Depth 10 m. Roger Steene.

B

H. guttavarius. Cayman I. Depth 12 m. Roger Steene.

C

H. guttavarius. St. Vincent. Depth 12 m. Roger Steene.

D

H. guttavarius. Aquarium. Length 7 cm. Patrick Louisy.

E

H. guttavarius. Aquarium. Length 12 cm.

F

G

H. guttavarius. Guadeloupe, Lesser Antilles. Depth 8 m. Length 8 cm. Patrick Louisy.

Golden Hamlet
Hypoplectrus gummigutta

Plectropoma gummigutta Poey, 1851. Havana, Cuba.

Appears to be mainly confined to the western Caribbean between Nicaragua and Cuba, rarely north of Cuba. Locally common in a few places, but generally regarded as being rare. Occurs in various reef habitats in a depth range from the shallows to over 30 m, depending on the locality. A beautifully coloured species in being entirely golden-yellow, except for its snout, which has black and blue markings. Length to 13 cm.

H. gummigutta. Cayman I. Depth 10 m. Length 85 mm. Roger Steene.

Yellow-belly Hamlet *Hypoplectrus aberrans*

Hypoplectrus aberrans Poey, 1868. Havana, Cuba.

Cuba, Lesser Antilles and Panama region. Shallow reef habitats in a depth range from the shallows to about 15 m. Mostly golden-yellow with a dark dorsal fin, the dark colour extending down to about lateral line and grading into the yellow below. Juveniles with a black peduncular spot. Length to 12 cm.

H. aberrans. St. Vincent, Lesser Antilles. Depth 15 m. **A** Scott Michael. **B** Roger Steene.

FAMILY GRAMMATIDAE

Western Atlantic percoid family. Comprises 2 genera and 12 small plesiopid-like species. Those in the genus *Gramma* are best known and often observed, but most species in the genus *Lipogramma* are secretive or live in deep water, often over 100 m, and are not well known. Only one species of *Lipogramma* occurs in shallower depths and is included here.

FAMILY GRAMMATIDAE, **GENUS** *Gramma* Poey, 1868

Neutral. Type species: *Gramma loreta* Poey, 1868. Comprises 4 small species confined to tropical Western Atlantic. Rocky reefs, usually seen hovering in caves or crevices and orientated with belly to the bottom, vertical walls or ceilings. Eyes large and mouth moderately small, perfectly suited for zooplankton floating past.

Reaching only about 10 cm maximum, they are great aquarium fishes and can be kept in relatively small set-ups. They require lots of hiding places and swim-through passages. Food should be varied and supplied on a regular basis.

Gramma loreto. Type-species for genus. Cayman Island. Depth 12 m. Roger Steene.

Black-cap Basslet *Gramma melacara*

Gramma melacara Böhlke & Randall, 1963. Bahamas.

Mainly found in the north-western and southern Caribbean region, and less common in the rest of the region or north to the Bahamas. Various reef habitats with caves and crevices at 20 m or more, reported to 130 m depth. Often swimming with its belly orientated to wall or ceiling. Easily identified by the mostly purple colour, and dark area over the back from tip of snout, lower edge running through lower part of eye and up to dorsal fin, ending as a submarginal stripe. Large individuals, possibly males, with yellow scribbles in dark area on head. Length to 10 cm.

A

B

C

G. melacara. Aquarium. Frank Schneidewind.

D

G. melacara. Swimming upside-down. Cozumel, Mexico. Depth 20 m. Length 10 cm. Roger Steene.

Royal Gramma *Gramma loreto*

Gramma loreto Poey, 1868. Cuba.

Mainly in the Caribbean region, and ranging north to the Bahamas. Various shallow protected reef habitats with caves and crevices to at least 60 m depth. Easily identified by the purple and yellow colouration. Has yellow or dark stripe from tip of snout through eye to end of head, a second one from below eye over preopercle. An eye-sized black spot anteriorly in dorsal fin. Length to 8 cm.

A

G. loreto. Aquarium. Length 8 cm.

B

G. loreto. Aquarium. Length 4 cm.

C

G. loreto. Cayman Island. Depth 12 m. Roger Steene.

D

G. loreto. Aquarium. Frank Schneidewind.

Brazilian Basslet
Gramma brasiliensis

Gramma brasiliensis
Sazima, Gasparini & Moura, 1998.
Espírito, Brazil.

North-eastern Brazil and offshore islands. Rocky and coral reef habitats from shallow protected habitats to deep water. Readily identified in the wild by the blue colour (in natural light) anteriorly and yellow posteriorly from about the anus. Has a small, but distinct pupil-sized black spot anteriorly on dorsal fin. Length usually to 12 cm, rarely 16 cm.

A *G. brasiliensis.* Arraial do Cabo, Brazil. D. 5 m. L. 4 cm. Carlos Ferreira.

G. brasiliensis. **B** Abrolhos Is., Brazil. Depth 10 m. Length 6 cm. Jerry Allen. **C** aquarium, from Brazil. Frank Schneidewind.

Yellow-cheek Basslet
Gramma linki

Gramma linki Starck & Colin, 1978.
Cozumel I., Mexico.

Mainly Caribbean Sea, ranging to the Bahamas, but often very deep and may have gone unnoticed in some adjacent areas. Occurs along steep walls and usually in the shade of caves or crevices. Usually at depths of 30 m or more and in some areas not seen shallower than 60 m. Body dark, bluish or purplish, scales with yellow centres and head mostly yellow with dark markings on lips and cheeks. Length usually to 75 mm, rarely to 10 cm.

G. linki. Cayman Island, Greater Antilles. Paul Humann.

Family Grammatidae, Genus *Lipogramma* Böhlke, 1960

Neutral. Type species: *Lipogramma anabantoides* Böhlke, 1960. Comprises 8 small species confined to tropical Western Atlantic. Rocky reefs, mostly very deep along walls and rarely observed underwater.

Like the closely related *Gramma,* they make great aquarium fishes. These basslets require lots of hiding places and swim-through passages. Food should be varied and supplied on a regular basis.

1 *Lipogramma anabantoides* Böhlke, 1960	Bahamas
2 *Lipogramma evides* Robins & Colins, 1979	Arrowsmith Bank
3 *Lipogramma flavescens* Gilmore & Jones, 1988	Bahamas
4 *Lipogramma klayi* Randall, 1963	Curaçao
5 *Lipogramma regium* Robins & Colins, 1979	Puerto Rico
6 *Lipogramma robinsi* Gilmore, 1997	Belize
7 *Lipogramma roseum* Robins & Colins, 1979	Colombia
8 *Lipogramma trilineatum* Randall, 1963	Curaçao

Three-line Basslet *Lipogramma trilineatum*

Gramma trilineatum. Böhlke, 1963. Bahamas.

Known from the Bahamas and South Florida, ranging into Caribbean Sea. Rocky reef, usually along vertical walls in caves or crevices in a depth range of about 20–50 m. A shy species occasionally darting out to grab zooplankton floating into reach. Readily identified by the yellow over the top of the head and back with irridescent blue lines, one from tip of snout to dorsal fin origins and others extending from behind eye, forming 3 lines over the head, explaining its name. A small species, length to 4 cm, but rarely longer than 3 cm.

L. trilineatum. Off Bimini, Bahamas. Paul Humann.

FAMILY PLESIOPIDAE

Indo-Pacific percoid family, highly diverse in shape and size, as well as in behaviour. The diversity of the plesiopids is evident by the large number of genera, versus number of species. Comprises 7 genera, several monotypic. Some have 2–4 species, and the only genus with more species is *Plesiops*, in which 17 species are recognised. The monotypic *Fraudella* is only known from trawling grounds in Qld, Australia, whilst *Steeneichthys* has 2 secretive small species that are less than 30 mm long. All genera are reef dwellers, 2 are planktivores that school close to reefs, whilst the others are mostly secretive, solitary, feeding primarily on crawling benthic invertebrates. Most diverse in morphology are the planktivores, *Trachinops* and *Assessor,* the former being very elongate with lots of dorsal-fin rays and filling a niche in subtemperate waters that is taken up by other groups in tropical waters, and the latter being a mouth brooder. Spawning of *Trachinops* and *Assessor* was observed in captivity. *Trachinops* produces its brood in a cluster of about 200 relatively large eggs that look much like those produced by mouth brooders, but are much too large to fit in the mouth. Instead, the male cares for the eggs in small holes below rocks after removing sand, and curls its body around them. *Assessor* males take the brood in their mouth (see genus). Other plesiopids have sticky eggs that are put against rocks or dead coral surfaces, and species such as *Paraplesiops* have been observed guarding eggs in the wild.

All species make great aquarium pets, but some species are only suitable for cold water and require cooling systems. Planktivores need regular feeding with live foods such as mysids and will lose condition if fed only once or twice a day. The tropical species are generally shy, however they do make interesting pets. Of course the best known is the Comet, which is easy to keep, successfully aquacultured, and always surprises visiting friends when it makes an occasional appearance.

Genera Contents

Family Plesiopidae, Genus *Calloplesiops* Fowler & Bean, 1930

Masculine. Type species: *Calloplesiops niveus* Fowler & Bean, 1930 (=*Plesiops altivelis* Steindachner, 1903). Monotypic. Widespread tropical Indo-West Pacific and Red Sea. A graceful fish with very large fins, usually held erect, making it look much larger than it really is. The large eye spot and line from where caudal and anal fins meet gives the impression of the posterior-end being a large head, whilst the real eye and head blend in with the body. It usually disappears head first, looking more like a pufferfish moving backwards.

Calloplesiops altivelis poses for ichthyologist Jerry Allen. Seeing this fish for many years, unable to get a picture, Jerry finally found a friendly one in Flores, Indonesia. It was almost too friendly and from the surface I watched Jerry pushing this fish back, to get some distance between it and his camera. I was out of air and out of film, but luckily my wife had film left in her camera and we got the shot that says it all!

A

Comet *Calloplesiops altivelis*

Plesiops altivelis Steindachner, 1903.
Nias I., off w. Sumatra, Indonesia.

Widespread tropical Indo-West Pacific and Red Sea. Coastal to outer reef habitats, typically in the back of far reaching crevices. Depth range 3–50 m, depending on locality. In sheltered localities often shallow. Readily identified by its shape and colour. Looks black with white spots when seen in caves. Spots vary in size, usually becoming more numerous and proportionally smaller with age. Can reach 16 cm in total length, but usually to 12 cm.

B

C. altivelis. Flores, Indonesia. Depth 25 m. Length 16 cm.

C

C. altivelis. Lembeh Strait, Sulawesi, Indonesia. Depth 18 m. Length 12 cm.

D

C. altivelis. Aquacultured juveniles. Helmut Debelius.

E

F

C. altivelis. Indonesia. **E** Flores. Depth 18 m. Length 12 cm. **F** Lembeh Strait, Sulawesi. Depth 7 m. Length 12 cm.

G

H

C. altivelis. **G** Maldives. Depth 25 m. Length 10 cm. 75 mm. **H** Egypt, Red Sea. Depth 10 m. Length 11 cm.

Family Plesiopidae, **Genus** *Paraplesiops* Bleeker, 1875

Masculine. Type species: *Plesiops bleekeri* Günther, 1861. Endemic Australian genus with 5 species restricted to subtropical and temperate reefs. Secretive fishes, usually seen in the back of low crevices or caves and some species are only seen at night. However, they appear to be diurnal. Large adults are less shy and sometimes easily approached, but whilst remaining in the safety of the reef. They are generally referred to as 'blue devils', due to the southern and eastern species (the only 2 species regularly seen) that are predominantly blue in colour and have large eyes that may perhaps look "evil" to some people. In reality, they appear to be rather friendly by nature, moving about slowly and gracefully, feeding on tiny prey that are usually picked from the bottom. Their excellent eye sight became apparent from individuals kept in an aquarium. When fed, they would see the tiny mysids from their hide-out some distance away, eventhough these were well camouflaged against the sand on the bottom. Judging by their relatively large mouth, one would expect that they would eat much larger prey, but when fed on cut scallops, a food they certainly like, only small bits were swallowed. Jaws have only tiny teeth that are set on patches along the front of the mouth. Diet comprises mostly small crustaceans and shelled molluscs.

Paraplesiops meleagris. Port Phillip Bay, Victoria. A large individual up close, showing the fine teeth on the lower jaw.

The various species are easily kept in captivity, but require good care and a quiet environment with few other fish, plenty of places to hide and temperatures below 20°C for the southern species. Unhappy fish go dark or may look washed out. They make excellent companions for other fishes, such as seahorses or seadragons, and are an added attraction in large displays, such as in public aquariums.

Paraplesiops meleagris. Victorian coast, Australia. Checking out a movement on the sand, a potential prey that may be buried.

The largest species can reach 40 cm and are amongst the largest in the family. When spearfishing was popular, the eastern species was often targeted, and, as with all reef fishes, this caused a rapid decline in numbers. The species is now protected by law and a permit is needed to collect them for aquarium use. Reproduction depends on breeding pairs on reefs and small numbers of eggs are laid on rock surfaces in narrow crevices (usually vertical) which are guarded by the parents. Little else is known, such as the duration of egg-development, larval stage etc. Juveniles, about 25 mm long, settle in the same habitats as the adults. Variations between different populations of the south coast species *Paraplesiops meleagris* suggests that there is little exchange between them, due to limited geographical movement of early stages.

1 *Paraplesiops alisonae* Hoese & Kuiter, 1984
2 *Paraplesiops bleekeri* Gunther, 1861
3 *Paraplesiops meleagris* Peters, 1869
4 *Paraplesiops poweri* Ogilby, 1908
5 *Paraplesiops sinclairi* Hutchins, 1987

Northern Blue-devil *Paraplesiops poweri*

Paraplesiops poweri Ogilby, 1908. .Moreton Bay.

Coastal southern Qld and northern New South Wales, eastern Australia. A common species in Moreton Bay, where on shallow reefs under plate corals and in crevices. Readily identified in the area by its 'cichlid-like' shape and blue spots on the head. Body with indistinct dusky bands that are more pronounced when small. Length to 18 cm.

P. poweri. Moreton Bay, Qld, Australia. **A** aquarium. **B** depth 5 m. Length 15 cm.

P. poweri. Moreton Bay, Qld, Australia. Depth 5 m. Length 15 cm.

Western Blue-devil *Paraplesiops sinclairi*

Paraplesiops sinclairi Hutchins, 1987.
Lucky Bay, east of Esperance, WA.

Known only from south-western Australia, between the Recherche Archipelago and Lancelin. A secretive species on rocky and coral reefs, where in the back of crevices or under plate corals to depths of at least 30 m. A dark species that blends in well against the background of its habitat. Almost uniformly bluish dark-grey, pectoral fins yellowish, and other fins with pale blue fin margins. Head sometimes with small blue spots or blue blotch on each gill-cover. Length to about 22 cm.

P. sinclairi. Lucky Bay (type-locality), WA. Depth 25 m. Length 20 cm.

Southern Blue-devil *Paraplesiops meleagris*

Paraplesiops meleagris Ogilby, 1908. Adelaide, SA.

Southern Australia from southern Victoria to the Houtman Abrolhos off Western Australia. Coastal reefs and rocky estuaries. Adults usually in depths over 15 m, juveniles often more shallow. In large caves, in protected coastal waters. Juveniles commonly share caves with adults. Large adults are often curious towards divers and the blue devils are amongst photographers' favourite subjects. Length to 33 cm.

A

P. meleagris. Lucky Bay, southern WA, Australia. Depth 25 m. L. 25 cm.

B

P. meleagris. aquarium. Melbourne. Length 40 mm.

C

D

P. meleagris. **C** Carnac I., Western Australia. Depth 7 m. Length 30 cm. **D** aquarium. Melbourne. Length 10 cm.

E

P. meleagris. Port Philip Bay, Victoria, Australia. Depth 15 m. Length 32 cm.

Eastern Blue-devil *Paraplesiops bleekeri*

Plesiops bleekeri Günther, 1861. No locality.

Eastern Australia from southern Queensland to southern New South Wales. Coastal reefs and islands, rocky reefs in clear and deep estuaries, mainly common south of Sydney. Lives primarily in the back of crevices, but adults can be approached closely where common. A beautiful species and easily recognised by its colour and shape. Length to about 40 cm.

Remarks: a protected species in NSW.

A *P. bleekeri*. Sydney, NSW, Australia. Length 40 mm.

B *P. bleekeri*. Subadult. Jervis Bay, NSW, Australia. Depth 10 m.

C

D

P. bleekeri. Jervis Bay, NSW, Australia. Depth 20 m. **C** length 55 mm. Brachiopods in background. **D** length 15 cm.

E *P. bleekeri*. Jervis Bay, NSW, Australia. Depth 5 m. Length 40 cm.

Alison's Blue-devil *Paraplesiops alisonae*

Paraplesiops alisonae Hoese & Kuiter, 1984. Victoria.

Bass Strait region of northern Tasmania and Victoria, ranging west to Victor Harbour, SA. Very secretive and usually seen only at night in crevices when using a torch. However, aquarium behaviour suggests that these fishes are diurnal. Dusky, with or without dusky bars, blue spots on head and blue lines in dorsal and anal fins. Males adorned with blue, yellow and orange. Smallest species in genus, length to about 15 cm.

Remarks: this species looks rather similar to *Plesiops*.

A *P. alisonae.* Westernport, Victoria. Depth 4 m. Length 75 mm.

B *P. alisonae.* Aquarium. Portsea, Victoria. Length 40 mm.

C

D

P. alisonae. Westernport, Victoria. Depth 4 m. **C** length 75 mm. **D** length 65 mm.

E *P. alisonae.* Male. Port Philip Bay, Victoria, Australia. Depth 20 m. Length 15 cm.

Family Plesiopidae, Genus *Plesiops* Oken, 1817

Masculine. Type species: *Plesiops nigricans* Rüppell, 1828. Indo-Pacific and Red Sea with 18 described species. Small and very secretive species that can be found in a variety of reef habitats, ranging from tidal pools to deep offshore reefs in excess of 100 m. Most species are drab in colour, but make excellent aquarium fish, especially in reef aquariums. They feed primarily on non-sessile invertebrates, but may take to small fishes as well.

n.i. = not included

White-spotted Longfin *Plesiops nigricans*

Pharopteryx nigricans Rüppell, 1828. Red Sea.

Red Sea and Gulf of Aden. Secretive in shallow rubble reefs, along edges or under large broken coral pieces. Generally dark grey to brownish with small white spots, sometimes showing faint barring on body and dark markings radiating from behind eyes. Length reported to 14 cm, but usually less than 12 cm.

A *P. nigricans*. Red Sea. Length 7 cm. Phil Heemstra.

B *P. nigricans*. Red Sea. Length 9 cm. H. Zetzsche.

Blue-spot Longfin *Plesiops corallicola*

Plesiops corallicola Bleeker, 1853. Priaman, Sumatra.

Indo Pacific, eastern Indian Ocean to central Pacific, and in the West Pacific ranging north to Taiwan, but appears to be absent in Australia and Coral Sea (reported in error). Secretive on shallow reef flats. Similar to *Plesiops nigricans*, but with blue spots and a blue-centred ocellus on cheek. Length to 15 cm.

A *P. corallicola*. Christmas I., I.O. Length 9 cm. Jerry Allen

B *P. corallicola*. Irian Jaya. Length 10 cm. Jerry Allen.

Tail-band Longfin *Plesiops oxycephalus*

Plesiops oxycephalus Bleeker, 1855. Batu Is., Indonesia.

Widespread tropical Western Pacific, Indonesia, ranging to West Sumatra, Indian Ocean, southern Japan and to Fijian region of the Central Pacific. Secretive on sheltered reefs, usually found in amongst rubble on shallow reef flats, gutters or lagoons. Generally dark-brown with bluish shine. Identified by yellow to orange band in caudal fin. Some adults with large yellow area on operculum, possibly a male colouration. Length to 9 cm.

P. oxycephalus. After Bleeker, 1877.

White-based Longfin *Plesiops verecundus*

Plesiops verecundus Mooi, 1995. Fiji Islands.

Mainly Western Pacific, ranging from Sumatra and western Australia to the Philippines and to Fijian Islands. Sheltered shallow reef flats and lagoons amongst coral rubble, usually at depths of a few metres, but reported to 30 m. Some geographical variations. The Fijian and true species has a yellow to reddish caudal fin with bands, whilst in the western Australian form the fin is dusky and plain. Similar to *Plesiops coeruleolineatus* in having orange tips on the dorsal-fin spines, but differes in having broad pale bands along the base of the dorsal and anal fins. Length to about 11 cm.

A

P. verecundus. Fiji. Holotype. 84 mm. Rick Winterbottom.

B

P. verecundus. Exmouth, Australia. D. 3 m. L. 45 mm. Jerry Allen.

C

P. verecundus. Fiji. Length 78 mm. Rick Winterbottom.

Coral Sea Longfin *Plesiops insularis*

Plesiops insularis Mooi & Randall, 1991. Lord Howe Island.

Known only from Coral and Tasman Seas region. Secretive on shallow reef flats, in lagoons and rubble slopes down to a depth of at least 25 m. Body generally dusky brown, head mostly yellow on sides. Dorsal and anal fin yellowish with pale blue longitudinal lines. Caudal fin mostly yellow and ventral fins pale with yellow tips. Length to about 11 cm

P. insularis. New Caledonia. Length 86 mm. Rick Winterbottom.

Blue-line Longfin *Plesiops coeruleolineatus*

Plesiops coeruleolineatus Rüppell, 1835. Red Sea.

Red Sea species. Secretive on shallow reef flats. Body generally dark, plain dusky brown. Pectoral fin transparent and its base bluish. Dorsal fin coloured as body with pale blue longitudinal line and yellow-orange spine tips. Soft-rayed part of dorsal fin, caudal and anal fin pale with broad yellowish margin and eye yellowish. Ventral fins pale without spotting. Length to about 10 cm

P. coeruleolineatus. Red Sea. Length 55 mm. Phil Heemstra.

Orange-lined Longfin *Plesiops* cf *coeruleolineatus*

Undetermined species.

Several forms referred to as *Plesiops coeruleolineatus* appear to represent different species. **A** has several orange lines in the dorsal and anal fin and a yellow submarginal stripe posteriorly in the caudal fin. The dorsal spine-tips are pointed as in *Plesiops melas*. This form is also known from Micronesia. **B** & **C** have similarly coloured dorsal and anal fins, but the dorsal spine-tips are squared off like in *Plesiops verecundus*. The submarginal yellow stripe in the caudal fin is placed more anteriorly and they have different head patterns. These fish need further investigation. Length to about 80 mm.

A

P. cf *coeruleolineatus*. New Caledonia. L. 58 mm. Rick Winterbottom.

B

C

P. cf *caeruleolineatus*. **B** Phuket, Thailand. Length 45 mm. **C** Comores. Length 33 mm. Rick Winterbottom.

Crimson-tipped Longfin *Plesiops melas*

Plesiops melas Bleeker, 1849. Bali, Indonesia.

Appears to be widespread West-Pacific. Most commonly found species in genus in Indonesia, juveniles often in tide-pools under large rubble pieces, but rarely seen during the day, unless looking under rubble in tide pools. Coastal to outer reef flats and lagoons, and reported to about 20 m depth. General appearance black. Dorsal and anal fin with blue line and dorsal with pointed orange-red spine tips. Eye blue, and a moderately large pale or white spot situated centrally on pectoral-fin base. May show a banded pattern and white spots on the head at night. Length to 75 mm.

A

P. melas. Tide pool, Flores, Indonesia. Length 50 mm.

B

C

P. melas. Tide pool, Flores, Indonesia. **B** night colour (same individual as in **A**). Length 50 mm. **C** length 30 mm.

A

P. mystaxus. Comores. Holotype, length 68 mm. Rick Winterbottom.

Moustache Longfin *Plesiops mystaxus*

Plesiops mystaxus Mooi, 1995. Comores Archipelago.

Known from the Comores, Madagascar, Red Sea to Gulf of Aqaba and Arabian Sea. Secretive on shallow reef flats. Collected as deep as 8 m. Adults have a black line along the top of the mouth that may be present in other species, but is less distinct. Dorsal-fin spine with pointed orange tips. Head with pale blue spots that may fade quickly after preservation. Length to 9 cm.

B

C

P. mystaxus. Comores. **B** length 42 mm. **C** juvenile, length 21 mm. Rick Winterbottom.

Andaman Longfin *Plesiops auritus*

Plesiops auritus Mooi, 1995. Sri Lanka.

Andaman Sea, Sri Lanka to Phuket, Thailand, and ranging into Indonesia to Java. Shallow reef flats amongst scattered coral boulders and rubble, usually in a few metres depth, but reported to 15 m. Named for dark 'ear' spot at upper edge of opercle, but this is not always distinct, especially when juvenile. Spine-tips of dorsal-fin yellow in adults and a blue line runs just below bottom of the incised membranes. Length to about 9 cm.

A

B

P. auritus. Phuket, Thailand. Lengths: **A** 45 mm. **B** 24 mm. Rick Winterbottom.

Yellow-banded Longfin *Plesiops* sp 1

Undetermined species.

Specimen from the Philippines that could not be identified at this stage. Going by the most recent revision by Mooi, 1995 it would probably be included with *Plesiops caeruleolineatus*,. However, in my opinion it seems clear that a number of species are going under this name and the true species may be restricted to the Red Sea region. Length to at least 60 mm.

P. sp 1. Philippines. Length 60 mm. Rick Winterbottom.

FAMILY PLESIOPIDAE, **GENUS** *Steeneichthys* Allen & Randall, 1985

Masculine. Type species: *Steeneichthys plesiopsus* Allen & Randall, 1985. Comprises 2 described species from the Indo Pacific. Small and secretive and it is likely more species will be discovered. They have only nine dorsal-fin spines: seven soft rays in anal fin, and like all plesiopids, 1 spine and 4 rays in the ventral fins but these are exceptionally long.

Roger's Pygmy-longfin *Steeneichthys plesiopsus*

Steeneichthys plesiopsus Allen & Randall, 1985.
Rowley Shoals, off north-western Australia.

Known only from a few scientific collections conducted in islands off north-western Australia, Samoa and Fiji. Occurs on slopes adjacent to drop-offs amongst rubble at depths from 15 to 30 m. Length to about 30 mm.

S. plesiopsus. Madang, PNG. D. 16 m. L. 28 mm. Jerry Allen.

Black Pygmy-longfin *Steeneichthys nativiatis*

Steeneichthys nativiatis Allen, 1987. Christmas I., Indian Ocean.

Known only from a single specimen from Christmas Island, Indian Ocean. It was collected in a rotenone station at a depth between 30–40 m at the bottom of a vertical cliff. It is likely that it will be found in other Indian Ocean localities. The specimen measures 24.5 mm SL.

S. nativiatis. Christmas I., Indian Ocean. D. 30 m. L. 30 mm.

FAMILY PLESIOPIDAE, **GENUS** *Fraudella* Whitley, 1935

Feminine. Type species: *Fraudella carassiops* Whitley, 1935. Monotypic genus from Queensland, Australia. A small species with a strongly serrated preopercular margin (smooth in other plesiopids) and usually 13 dorsal spines. An interesting and colourful little species that would probably be great in a reef aquarium, and one that can probably be aquacultured with relative ease. May employ parental care of brood, in either a shell or orally.

Red Pygmy-devil *Fraudella carassiops*

Fraudella carassiops Whitley, 1935. Qld.

Mainly known from the Capricorn-Bunker group inside the Great Barrier reef, Queensland, Australia. Primarily known from scallop grounds where trawled in depths between 28 and 50 m. Nothing is known about its behaviour. Judging by its size and colour, one would expect that it lives secretively in objects out on open substrates, perhaps with sponges, or uses scallop shells and other mollusk shells as a hide-out. Colour bright red with orange caudal fin and soft part of dorsal fin. Colour quickly fades after capture. Length to about 6 cm.

F. carassiops. Trawled, Queensland, Australia. Length 35 mm. Darrin Roy.

FAMILY PLESIOPIDAE, GENUS *Trachinops* Günther, 1861

Masculine. Type species: *Trachinops taeniatus* Günther, 1861. Endemic Australian genus with 4 species restricted to subtemperate and temperate reefs. They are doubtfully included with the Plesiopidae and certainly are unusual amongst these, being schooling planktivores rather than the otherwise cryptic reef dwellers. Mostly small fishes, just exceeding 10 cm in length, they form schools that may fill the water column from the bottom to the surface in some areas, with hundreds and sometimes thousands of individuals. They are most numerous in the mouth of large open estuaries along rocky reefs with boulders or walls where tidal currents sweep past. During the slack they often retreat to the reefs and may rest in caves or swim slowly in large groups nearby. These fishes fill a niche that interestingly is filled by species of other families in different areas. For example, in New Zealand there are the tripterygiids, a type of blenny that normally hugs the bottom, that has adapted to feeding on small zooplankton in temperate waters, also becoming more elongate than its closely related bottom dwelling siblings. Similarly, in Japan a number of gobies have taken on this role.

Trachinops noarlungae. Yawning or stretching its muscles. As planktivores, they typically have a proportionally large and protrusible mouth to suck-in prey in open water.

Trachinops taeniatus. Male guarding eggs in aquarium. It removes sand from under rocks to created a small burrow, into which it just fits with its body in a curved position around the egg-mass.

Large individuals normally have long pointed caudal fins that become greatly extended in the centre with age. When swimming it has a sideways, undulating movement which gave rise to its popular name 'hulafish'.

The various species are easily kept in captivity, but require regular feeding of small prey such as mysids. However, they will take to a variety of meat-based foods. The problem is regular supply as these fishes normally feed most of the day on small zooplankton. In addition, temperatures need to be below 20°C for the southern species and the eastern and most northern species will not tolerate it much higher either. If these conditions are right, they are marvellous fishes that do best when kept in small groups of 10 or more individuals. They will readily breed. *Trachinops taeniatus* was kept in a 2 m long aquarium with a variety of gobies and syngnathids and eggs were produced during the summer on a monthly basis. The hatching tiny larvae, just a few millimetres long, were attracted to light and obviously pelagic, but no attempt was made to grow them.

The largest species can reach 15 cm in Tasmania, where temperatures are lowest, but most reach about 10 cm in length, excluding the caudal fin extensions.

1 *Trachinops brauni* Allen, 1977
2 *Trachinops caudimaculatus* McCoy, 1890
3 *Trachinops noarlungae* Glover, 1974
4 *Trachinops taeniatus* Günther, 1861

Eastern Hulafish *Trachinops taeniatus*

Trachinops taeniatus Günther, 1861. New South Wales.

A common NSW species, just ranging into Queensland in the north and Victoria in the south. Occurs commonly in rocky estuaries and coastal reefs. Usually forms large schools on coastal reefs to about 20 m depth, but may rise high above to feed on zooplankton during tidal currents. Easily identified by its shape and the distinctive multi colouration. Males develop long lanceolate caudal fins. Length to 10 cm.

A

P. taeniatus. Aquarium. Sydney specimens. Lengths ~10 cm.

B

C

P. taeniatus. Sydney, Australia. Depth ~ 5 m. **B** yawning. Length ~8 cm. **C** unusual yellow back.

D

P. taeniatus. Sydney, Australia. Length 10 cm.

E

P. taeniatus. Schooling. Bermagui, NSW, Australia.

Southern Hulafish *Trachinops caudimaculatus*

Trachinops caudimaculatus McCoy, 1890. Victoria.

Tasmania, southern Victoria, and ranging west to islands in the Great Australian Bight of South Australia. Particularly common in Tasmania, such as in the Derwent Estuary where often in enormous schools feeding high above the bottom in pursuit of zooplankton. Various habitats from shallow piers to offshore rocky reefs - anywhere that offers crevices they can get into. Length to 15 cm.

A

T. caudimaculatus. Victoria, Australia. Depth 4 m. Length ~10 cm.

B

C

T. caudimaculatus. Victoria, Australia. Depth 15 m. Lengths 80–95 mm.

Braun's Hulafish *Trachinops brauni*

Trachinops brauni Allen, 1977. Rottnest I., WA.

Restricted to the south-west corner of Australia. Sheltered coastal bays on rocky or coral reefs to about 35 m depth. Usually occurs in small aggregations that stay close to reefs and wait for zooplankton to come into close reach. Readily identified by its black and bright blue longitudinal stripes and its elongate shape. Length to 80 mm.

A

B

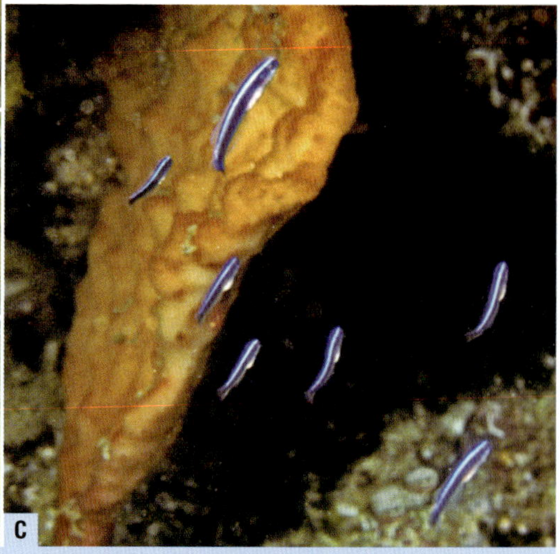

C

T. brauni. Western Australia. **A** & **B** Carnac I. Depth 10 m. Length 8 cm. **C** Mondrain I. Depth 20 m. Lengths 35–40 mm.

Noarlunga Hulafish *Trachinops noarlungae*

Trachinops noarlungae Glover, 1974. South Australia.

Australia's south coast from Kangaroo Island, South Australia, to the Houtman Abrolhos off Western Australia. Coastal bays and offshore reefs. A common species in the southern parts of the S.A. gulfs and islands off the south coast where usually in large aggregations. Juveniles and adults often mix. Identified by its pale colour with yellow on head and tail, males with bright blue margins on caudal fin and long central extension. Length to 15 cm, including long caudal fin.

A

B

C

T. noarlungae. Schooling adults, mixed with juveniles. Pearson Island, Great Australian Bight. Depth 15–20 m. **C** length 14 cm.

D

T. noarlungae. Noarlunga, South Australia (type-locality), Australia. Depth 5 m. Length 12 cm.

E

T. noarlungae. Schooling adults. Top Gallant Island, Great Australian Bight. Depth 25 m. Length 12–14 cm.

Family Plesiopidae, Genus *Assessor* Whitley, 1935

Masculine. Type species: *Assessor macneilli* Whitley, 1935. A western Pacific genus comprising 3 species. Small fishes, reaching just 60 mm, that typically swim upside-down on the ceiling or back of large caves or reef-overhangs. They feed on zooplankton that is carried by currents sweeping reefs. The caudal fin is forked with broad lobes.

Two specimens of *Assessor macneilli* were collected and kept for about 4 months in a small aquarium with a few companion fish that included a pair of *Nemateleotris magnifica*, a single *Belonepterygion fasciolatum* and a juvenile *Canthigaster ocellicincta*. It was only then that it was discovered that these Scissortails are mouth-brooders. The male incubates the brood, comprising about 200 eggs for 15–16 days and during this time it swims out in the open and behaves more or less normally, except it does not appear to feed. For several days prior to hatching, the male keeps its mouth open and the eggs are shifted back and forth at intervals of between 2 and 8 minutes. Hatching was not witnessed and evidently takes place at night as fry were usually discovered in the early morning. The larvae, about 4–5 mm long, swim close to the surface and are mostly transparent. Most usually disappeared after a few hours due to the filtering system. A new brood would appear in one or two days after hatching and this cycle was repeated 5 times.

A

A. flavissimus. Lizard I., Qld, Australia. Depth 10 m. Length 60 mm.

Yellow Scissortail *Assessor flavissimus*

Assessor flavissimus Allen & Kuiter, 1976. Qld.

Northern Great Barrier Reef and south-eastern Papua New Guinea. Occurs in small groups and typically swim upside-down in caves. Mainly occurs in inner and outer reef habitats that offer protection during heavy seas. Often found mixed with its blue sibling *Assessor macneilli*. This species is readily identified by its shape and bright yellow colour. In deeper water it often appears greenish, especially in PNG. Length to 60 mm.

B

A. flavissimus. Lizard I., Qld, Australia. Length 60 mm.

C

A. flavissimus. Milne Bay, PNG. Depth 6 m. Length 50 mm.

D

A. flavissimus. Lizard I., Qld, Australia. Length 60 mm.

Blue Scissortail *Assessor macneilli*

Assessor macneilli Whitley, 1935. Queensland.

Great Barrier Reef and Coral Sea. Coastal to inner reefs, usually in sheltered parts of reef channels where in schools in large caves or below overhangs. Typically upside-down in caves or with belly towards vertical walls. When on the bottom of an aquarium it usually swims the right way up. Spawns in captivity when given the right conditions (see genus text). Identified by the dark blue colour, shape and behaviour. Length to 60 mm.

A. macneilli. Qld, Australia. **A** & **B** aquarium, latter showing brood. **C** Lizard I. Depth 15 m. Length 60 mm.

Japanese Scissortail *Assessor randalli*

Assessor randalli Allen & Kuiter, 1976. Japan.

Southern Japan, Kerama and Ryukyu Islands, and south to Taiwan. Mainly in large shaded caves on well-protected inner reefs. Usually occurs singly or in small groups and appears to be uncommon, but is easily overlooked due to its small size, dark colour, and somewhat secretive behaviour in caves. Dark blue overall and yellow spotting on median fins, and recognised best by its shape and behaviour. Length to 50 mm.

A. randalli. Kerama I., Japan. Depth 10 m. Length 50 mm.

Family Acanthoclinidae

A small family of mostly tiny fishes with 4 genera and about 10 known species, but more are likely to be discovered. Closely related to Plesiopidae and regarded by some taxonomists as a subfamily of it. Most genera comprise species that reach just a few centimetres in total length. They live secretively in rocky reefs or coral rubble and remain hidden under cover. They are small but fierce predators on postlarval reef fishes. They are similar to dottybacks and plesiopids, but feature mostly spines in the dorsal and anal fins. They are also known as dottybacks, rockfishes or spiny-basslets.

Easily maintained in the aquarium, but *Acanthoclinus* grows moderately large and lives in cold water. Male looks after brood in holes excavated under rocks. Hatching after 1–2 weeks and young pelagic, settling on reefs at small size (5–10 mm).

Genus *Acanthoclinus* McCulloch, 1915

Masculine. Type species: *Acanthoclinus fuscus* Jenyns. Appears to be restricted to the New Zealand region with 5 species. The common *A. fuscus* is included here. *A. littoreus* (Forster, 1801), *A. marilynae, A. matti, & A. rua* (all Hardy, 1985) are far less common. They feature 3 lateral lines running the full length of the body. Most are found in shallow coastal waters under rocks. Lengths vary from about 15–30 cm, depending on the species.

A. fuscus. Aquarium, New Zealand. Length 12 cm.

Olive Rock-fish *Acanthoclinus fuscus*

Acanthoclinus fuscus Jenyns, 1842. Bay of Islands, New Zealand.

Endemic to New Zealand from Cape Reinga to Steward Island. Coastal reefs, rockpools and estuaries. Secretive, usually amongst boulders or under rocks. Dark brown to greenish brown with a blotched pattern of grey to black spots and a few white spots scattered over the body, together with a white nape stripe from tip of snout to dorsal fin origin. Length to 30 cm.

Genus *Acanthoplesiops* Regan, 1912

Masculine. Type species: *Acanthoclinus indicus* Day, 1888. Small Indo-Pacific genus with 4 tiny species. Extremely cryptic and usually collected from under rocks, or when using chemicals. Rarely photographed and 2 of the species are included. Besides those, and the type-species from India, Hardy, 1984, described *A. psilogaster* from Taiwan.

A. hiatti. Flores, Indonesia. Depth 15 m. Length 20 mm.

Skunk Spiny-basslet *Acanthoplesiops hiatti*

Acanthoplesiops hiatti Schultz, 1953. Off Eniwetok, w. Pacific.

Widespread Indo-West Pacific, known mainly from Indonesia and Philippines but also in Fiji. Sheltered rubble reef slopes to at least 20 m depth. Nearly all black, but with a thin white line over the back from the tip of the snout and along the tips of the dorsal fin spines. Length to 24 mm.

A. echinatus. Sulawesi, Indonesia. Length 30 mm. G.R. ALLEN.

Badger Spiny-basslet *Acanthoplesiops echinatus*

Acanthoclinus echinatus Smith-Vaniz & Johnson, 1990. Ambon.

Moluccan Seas, Ambon south to Flores and west to Togean Bay, Sulawesi. Secretive in sheltered reefs to about 20 m depth. Pale cream with white blotches and distinctively marked with black banding over the top, anal fin and a vertical band on the caudal fin base. Length to 40 mm.

Genus *Belonepterygion* McCulloch, 1915

Neutral. Type species: *Acanthoclinus fasciolatus* Ogilby, 1889. Monotypic genus. Small, secretive, Indo-Pacific.

Excellent in small aquariums, usually feeding on small crawling or swimming invertebrates, but may prey on small fishes as well. Shy at first.

Barred Spiny-basslet *Belonepterygion fasciolatum*

Acanthoclinus fasciolatus Ogilby, 1889. Lord Howe I.

Widespread Indo-West Pacific, tropical Australian waters to southern Japan. Usually found amongst large coral rubble patches in lagoons or tidal channels, usually in less than 5 m depth. Very secretive, usually staying in the shade. A white line over the top of the head and usually the body with thin vertical black lines or a white mid-lateral line on the head and anterior part of body. Length to 50 mm.

A

B. fasciolatum. Moreton Bay, Qld. D. 3 m. L. 50 mm. Tom Bowling.

B

C

B. fasciolatum. One Tree Island, Qld, Australia. Depth 4 m. Lengths: **B** 35 mm. **C** 45 mm.

Genus *Beliops* Hardy, 1984

Masculine. Type species: *Beliops xanthokrossos* Hardy, 1984. Monotypic genus. Small, secretive, temperate Australia.

One specimen was placed in a small aquarium with plenty of rocks and a few other fishes, but it never showed itself.

Southern Spiny-basslet *Beliops xanthokrossos*

Beliops xanthokrossos Hardy, 1984. Lord Howe I.

Australia's south coast. Lives under rubble, rocks or debris on semi-open substrate. May look black with just a white line over the nape from tip of snout to dorsal-fin origin, but sometimes yellowish with black spots and scattered small white spots. Dorsal fin tips yellow. Length to 26 mm.

A

B. xanthokrossos. Edithburg, S.A., Australia. D. 12 m. L. 20 mm.

BIBLIOGRAPHY

Allen, Gerald R., Ross Robertson, 1994. *Fishes of the Tropical Eastern Pacific*. Crawford House Press, Bathurst, New South Wales.

Carvalho-Filho, Alfredo, 1999. *Peixes: costa brasileira.* Melro, São Paulo.

Debelius, Helmut, 1997. *Mediterranean and Atlantic Fish Guide.* IKAN-Unterwasserarchiv, Germany.

Debelius, Helmut, 1998. *Red Sea Reef Guide.* IKAN-Unterwasserarchiv, Germany.

Edwards, Alasdair, 1990. Fish and Fisheries of Saint Helena Island. Gov. St. Helena.

Francis, Malcolm, 2001. *Coastal Fishes of New Zealand.* An identification guide, third edition. Reed Publishing, Auckland.

Gloerfelt-Tarp, Thomas & Patricia J. Kailola. *Trawled Fishes of Southern Indonesia and Northwestern Australia.*

Humann, Paul, 1993. *Reef Fish Identification, Galápagos.* New World Publications, Inc., Jacksonville.

Humann, Paul, 1997. *Fisch Führer Karibik.* Jhar Verlag, Hamburg.

King, Dennis, 1997. *Reef Fishes & Corals.* Struik Publishers P/L, Cape Town.

King, Dennis, & Valda Fraser, 2002. *More Reef Fishes & Nudibranchs.* Struik Publishers P/L, Cape Town.

Kuiter, Rudie H., 1992. *Tropical Reef-Fishes of the Western Pacific Indonesia and Adjacent Waters*. Penerbit PT Gramedia Pustaka Utama, Jakarta.

Kuiter, Rudie H., 1996. *Guide to Sea Fishes of Australia*. New Holland Publishers, Australia.

Kuiter, Rudie H., 1998. *Photo Guide to Fishes of the Maldives*. Atoll Editions, Apollo Bay.

Kuiter, Rudie H., 2000. *Coastal Fishes of South-Eastern Australia*. Gary Allen Pty Ltd. Sydney, Australia.

Kuiter, Rudie H., & Helmut Debelius, 1994. *Southeast Asia Tropical Fish Guide*. IKAN-Unterwasserarchiv, Germany.

Kuiter, Rudie H., & Takamasa Tonozuka, 2004. *Indonesian Reef Fishes*. PT Dive & Dive's. Bali. 3 volumes.

Louisy, Patrick, 2002. *Guide d'indentification des Poissons Marins*. Ulmer, Paris.

Masuda, Hajime, K. Amaoka, C. Araga, T. Uyeno & T. Yoshino, 1984. *The Fishes of the Japanese Archipelago.* Tokai University Press, Tokyo.

Merlen, Godfrey, 1988. *A Field Guide to the Fishes of Galápagos.* Wilmot Books, London.

Michael, Scott W., 1998. *Reef Fishes Volume 1.* Microcosm, London.

Myers, Robert F., 1999. *Micronesian Reef Fishes.* Coral Graphics, Guam.

Nakabo, Tetsuji, 1993. *Fishes of Japan with Pictorial keys to The Species.* Tokai University Press, Tokyo.

Nakabo, Tetsuji, 2002. *Fishes of Japan with Pictorial keys to The Species, English edition.* Tokai Un. Press, Tokyo.

Randall, John E., 1996. *Shore Fishes of Hawai'i.* Natural World. Press, Vida, Origan.

Smith, Margaret M. & Phillip C. Heemstra, 1986. *Smith's Seafishes.* Springer-Verlag.

OTHER BOOKS IN THIS SERIES....

Angelfishes
A Comprehensive Guide to Pomacanthidae
by Helmut Debelius, Hiroyuki Tanaka & Rudie H Kuiter

• 208 pages • Hardback
• ISBN 0-9539097-5-1

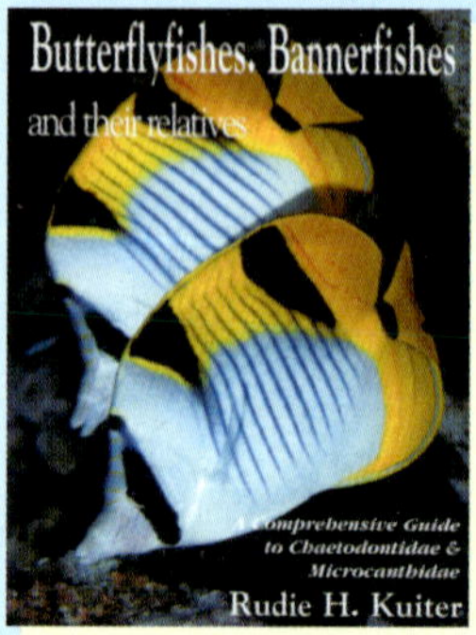

Butterflyfishes, Bannerfishes and Their Relatives
A Comprehensive Guide to Chaetodontidae & Microcanthidae
by Rudie H Kuiter

• 208 pages • Hardback
• ISBN 0-9539097-3-5

Fairy & Rainbow Wrasses and Their Relatives
A Comprehensive Guide to Selected Labroids
by Rudie H Kuiter

• 208 pages • Hardback
• ISBN 0-9539097-2-7

Surgeonfishes, Rabbitfishes and Their Relatives
A Comprehensive Guide to Acanthuroidei
by Rudie H Kuiter & Helmut Debelius

• 208 pages • Hardback.
• ISBN 0-9539097-1-9

Seahorses, Pipefishes and Their Relatives - A Comprehensive Guide to Syngnathiformes
by Rudie H Kuiter

• 240 pages • Hardback
• ISBN 0-9539097-4-3

Marine Fish Families Box Set Volume 1

Contains:-
- Seahorses, Pipefishes & Their Relatives
- Surgeonfishes, Rabbitfishes & Their Relatives
- Fairy & Rainbow Wrasses & Their Relatives

All available on-line at www.tmc-publishing.com